AF553674

Radiation and Climate Change

Radiation and Climate Change

Authors
Gulshan Dhingra
Sanjeev Lal

2013
BIOTECH BOOKS®

ISBN 978-81-7622-272-3

Published by: **BIOTECH BOOKS®**
4762-63/23, Ansari Road, Darya Ganj,
New Delhi - 110 002
Phone: +91-011-23262132
E-mail: biotechbooks@yahoo.co.in

Printed at: **Chawla Offset Printers**
Delhi - 110 052

PRINTED IN INDIA

Preface

The Sun is the most important external driver of climate. Although the total solar irradiance (TSI) fluctuates by less than 0.1 % during a solar cycle, the solar impact on the terrestrial atmosphere can be significant, in particular in the upper atmosphere where the highly variable energetic part of the solar spectrum is absorbed. TSI variations on centennial time scales are of similar magnitude. Unfortunately, the scientific understanding of the variation of solar radiation (and its spectral components) and its impact on the atmosphere is rather limited. This concerns the direct modification of composition by solar radiation, but even more so various coupling mechanisms. For example, photo chemically active gases are generated in the upper atmosphere; propagate to lower layers where they substantially alter the composition, e.g., the abundance of ozone. The Ultraviolet portion of the sunlight (UV-B) is detrimental to the plant and animal cell. The biological effects of UV-B radiation, particularly, UV-B portion of spectrum (280-315 nm), which is associated with stratospheric ozone (O_3) depletion, may exert effects on the terrestrial ecosystem, through action on different organisms and their life processes.

Most important consequences will be indirect effects of elevated UV-B radiation, causing changes in the chemical composition or form of organism. The changes can result in either a decrease or an increase in susceptibility. Plants like other organisms, sense and respond to UV-B radiation, but detrimental effects are only observed , when levels of biologically active UV-B radiation exceeds the plant's natural capacity to avoid exposure and defend itself. As can be seen in various chapters in this book, major progress has been achieved in our understanding of solar radiation, its impact on the plants especially Indian Mustard and the role of some plant growth regulators to mitigate or to overcome these hazardous impacts.

The purpose of this book is to summarize the basic concept of electromagnetic radiation coming from the sun and how UV-B radiation affects plants at morphological, anatomical and biochemical levels. Authors have contributed their findings of various deleterious effects of UV-B on two varieties of *Brassica* sps (Mustard plant) with relevant data and graphs. The book also describes the role of plant growth hormones at various levels and the important role played by these PGRs to overcome the negative impacts caused by ultraviolet radiation. We earnestly believe that the book will be a valuable kit for postgraduate and research students.

The authors are grateful to Dr. V.K. Jain, Principal, Govt. Degree College, Aanwla (Bareilly) for his guidance and untiring support. We acknowledge the encouragement provided by Dr. Gagan Matta, Gurukul Kangari University, Haridwar and support of Prof. D.R. Khanna for completion of this book. Finally, we thank Mrs Neetu Dhingra, Muskaan and Prachi for their understanding and support throughout the development of this book.

Errors and inaccuracies if any will be corrected through feedback and suggestions from readers.

Gulshan Dhingra

Sanjeev Lal

Contents

Chapter 1
Introduction

The ultraviolet radiation (UV) is usually, defined as electromagnetic radiation between the shorter wavelength of X- rays and longer wavelength of visible radiation. The ultraviolet portion of the sunlight (UV-B) is detrimental to the plant and animal cell. The different wavelength of the electromagnetic radiation, cause deleterious impacts on plants, animals as well as on the human beings. The UV-B radiation is divided into three wavelengths and their ranges are as UV-A, UV-B and UV-C.

The biological effects of UV-B radiation particularly, UV- B portion of the spectrum (280-315 nm), which is associated with stratospheric ozone (O_3) depletion, may exert effects on the terrestrial ecosystem, through action on different organism and their life processes. Most important consequences will be indirect effects of elevated UV-B radiation, causing changes in the chemical composition or form of organism. The changes can result in either a decrease or an increase in susceptibility. Plants like other organisms, sense and respond to UV-B radiation, but detrimental effects are only observed, when levels of biologically active UV-B radiation exceeds the plants, natural capacity to avoid exposure and defend itself (Neeta Bhatt *et al.*, 2004).

The amount of UV-B radiation, reaching the surface of earth is highly variable as it is influenced by many factors. The solar zenith angle and day length, determined by season and latitude, account for the majority of variation (Webb *et al.*, 1997). Ozone (O_3) is the primary UV-B absorbing component of the atmosphere with approximately 90 per cent of atmospheric ozone (O_3) column being in the stratosphere (ozone layer) and remainder in the troposphere (WMO, 1995). The amount of O_3 in both of these atmospheric regions has a great deal of temporal and spatial variability (Webb, 1997). Clouds, aerosols and surface albedo are also significant factors in determining the UV-B irradiation at particular location and time. These factors tend to result in UV-B irradiances being highest in the summer at in lower latitudes and at high altitudes (Mandronich *et al.*, 1995).

Naturally, UV-B distribution has been significantly affected by anthropogenic activity, primarily through the release of man-made chlorine and bromine compounds (CFC_S and halons) and selected solvent that increased global warming. These undergo sunlight and induced photochemical changes and could alter natural balance of creative and destructive process and in the stratosphere every CFC molecule atom can destroy up to 100,000 ozone (O_3) molecules. The use of CFC_S in the late 20th century, mainly as refrigerants and spray pollutants, led to the destruction of stratospheric ozone (Molina and Rowland *et al.*, 1974). The major global concern occurred in the spring at the highest latitudes, when sun-light returns to the very cold stratospheric clouds in the polar vertices, which provide condition for the very rapid chlorine and bromine, catalyzed O_3 destruction (Sorg, 1996). It was further observed by Sorg in 1990 that, this has been most apparent over Antarctica, where total ozone in October is less than 40 per cent. The ozone (O_3) hole results in greater UV-B irradiation in the spring at Palmer station, Antarctica (64° N) than any time of year in San Diego, USA (32° N), despite much lower incident sunlight (Mandronich *et al.*, 1995). Whilst not being as dramatic as the Antarctic hole, significant spring ozone depletion has occurred over the Arctic particularly in the 1960 (Pyle, 1997).Therefore, maximum ozone loss over the Northern mid-latitudes, relative to the situation in 1960, is expected to be increased of 12-13 per cent in the winter/spring and 6-7 per cent in the summer/autumn, producing increase in UV-B of 15 per cent and 8 per cent respectively (WMO, 1995).

Ozone (O_3) depletion by anthropogenic gases has increased the atmospheric transmission of solar UV-B (280-315 nm). The consequences of enhanced UV-B levels on primary producers have grown dramatically over the past 20 years, but it has been hampered by UV-B exposures mimic ozone depletion. The photosynthesis is more sensitive to UV-B in phytoplankton than in terrestrial plants, probably owing to less effective screening in phytoplankton. Productivity of terrestrial plants is usually unaffected by enhanced UV-B, although reduced growth has been observed and may be increase in magnitude over successive years. Aquatic productivity is often compromised by short-term exposures to enhanced UV-B and long-term assessments are complicated by dynamic-nature of aquatic systems and by non-linear responses. The recent work examining UV-B affects on multiple tropics levels suggests that outcomes will be diverse and difficult to predict (Thomas A. Day *et al.*, 2000).

Increase in penetration of UV-B (280-315 nm) to terrestrial surface as a consequence of depletion of the stratospheric ozone layer has received much global concern. 10 per cent depletion in stratospheric ozone corresponds to a 20 per cent increase in the fluence of biologically damaging UV-B, which is likely to increase in future, if ozone damage is not checked (Baker and Allen, 1994). The enhanced UV-B radiation can deleteriously affect overall growth and biomass accumulation in the plant species (Tevini, 2000). The plants contain a large number of UV-B sensitive targets as nucleic acids, lipids, proteins and quinines (Jordan, 1996), which must be protected to ensure the normal growth and development of plants. Failure to do so may lead to alternations in the over all morphology and physiology of many plants exposed to UV-B. Therefore, UV-B exposure is one of the major factors, which responsible for the low productivity of crop plants and natural vegetation, so has become an increasing threat for agriculture.

Progresses in the understanding of the mechanisms that mediate these effects of ambient UV-B have been slow for the various reasons. The most of the information on the deleterious impacts of UV-B at the molecular level have been obtained in the UV-B exposure, with monochromatic or heavily unbalanced UV-B sources *i.e.* sources that produced unnaturally high UV-B to photosynthetically active radiation (PAR) ratios. Under the monochromatic UV-B, other photoreceptors can be activated such as phytochromes. The UV-B exposure has the potential to damage key macromolecules and cellular structures, particularly, when high doses of UV-B was used.

Firstly, full compliance with the "Montreal protocol" is uncertain (Greene, 1995) and secondly, anthropogenic emissions of green-house gases and other pollutants have the potential to change troposphere ozone (O_3) concentrations, cloud cover and stratospheric air temperatures resulting in altered UV-B irradiation. A recent analysis using a global climate model predicted that rising CO_2 and other green-house gases may cause stratospheric cooling and more stable vertices in winter, delaying recovery on ozone (O_3) concentrations for approximately 15 years after declines in CFCs emissions (Shindell *et al.*, 1998).

The doubling of current levels of atmospheric CO_2 increase in other radioactively active gases has led to predictions of increase in global air temperature and shifts in precipitation patterns. Additionally, stratospheric ozone depletion may result in an increased UV-B radiation incident at the Earth's surface in the same areas. Since these changes in the earth's atmosphere may have profound effects on vegetation. Elevated CO_2 increases photosynthesis and usually results in an increased biomass and yield. The magnitude of these increases and specific photosynthetic response depends on the plant species and are strongly influenced by other environmental factors including temperature, light level and availability of water and nutrients. While elevated CO_2 reduces transpiration and increases photosynthetic water use efficiency, increasing air temperature can results in greater water use, accelerated plant developmental rate and shortened growth duration. The studies on UV-B exposure have demonstrated a wide range of photobiological responses among plants with decrease in photosynthesis and plant growth among more sensitive species (J.T Baker *et al.*, 1994).

Patterns of environmental changes in the biosphere include concurrent and sequential combinations of the increasing CO_2 levels; long term changes are resulting mainly from stratospheric O_3 depletion, greater troposphere O_3 photochemical synthesis and increasing CO_2 emissions. Effects of selected combinations were evaluated in tomato (*Lycopersicum esculentum*) seedling, using sequential exposures to enhanced UV-B radiation and O_3 in differential CO_2 concentrations. Exposure to enhanced UV-B, increased leaf chlorophyll and UV-absorbing compounds, but decreased leaf area and shoot/root ratio. The O_3 (ozone) exposures generally inhibited growth and leaf photosynthesis and did not affect UV-absorbing compounds. The highest dose of O_3 eliminated the stimulating effect of CO_2 enrichment after ambient UV-B pre-exposure on leaf photosynthesis. Pre-exposure to enhanced UV-B mitigated O_3 damage to leaf photosynthesis at the elevated CO_2 (X. Hao *et al.*, 2000).

Before the evolution of organisms capable of photosynthesis and therefore, production of substantial amounts of oxygen (O_2), on the surface of earth was exposed to appreciable UV-B and even a bit of UV-C, along with UV-A. Once oxygen became abundant in the atmosphere the sun-light induced reactions that converts oxygen to ozone (O_3). The ozone absorbs the UV-C and some of UV-B radiation. Ozone layer is found between 19 and 30 km. above the ground level in the part of atmosphere called stratosphere. As the ozone (O_3) layer gets thinner in the stratosphere the protective filter activity of atmosphere is progressively reduced. The UV-B spectral band (280-315 nm) contributes less then 2 per cent of short wave photons received by the terrestrial organisms in the most natural environment, concerns about potential impacts of stratosphere contributed to spark interest in the studies of plant responses to enhanced UV-B levels, during the last two decades (Caldwell *et al.*, 2003). Due to disturbance of the thermal structure of atmosphere, probably causes changes in the atmospheric circulation. The life time of these ozone (O_3) destroying substance is very long and they may continue to deplete the ozone layer long. Although the use of most CFCs molecules has been passed out, ozone depletion is currently near or at its minimum (Mckenzie *et al.*, 2003). In the case of plants and other primary producers, ozone depletion has its greatest effects by increasing the atmospheric transmission of solar UV-B radiation.

The present investigation is being carried out on the deleterious impacts of UV-B radiation in the Brassicacae family, also known as the Cruciferae family. The two selected varieties of mustard crops such as, *Brassica compestris PT-330* (Brown Sarson) and *Brassica juncea PR-15* (Rai), were investigated after exposure to artificial UV-B, alone and along with different concentrations of plant growth regulators such as IAA (10^{-7} M), Kn (10^{-5} M) and GA_3 (10^{-6} M) in *Brassica compestris PT-303* and (10^{-7} M) in *Brassica juncea PR-15* were applied. Because Indian mustard is a very promising species and is playing an even more important role in global agriculture and the horticulture, which has the good production, so there is an urgent need for the modern and economical development. Vegetable oils and seeds are the most important group of agricultural commodities in terms of value in the world trade. The mustard varieties of crop play an important role in the international market.

The importance of these crops is further enhanced by their wide adaptation. The mustard crop is cultivated as winter crop in India, Pakistan, and Bangladesh and also cultivated in a Europe, Africa and Asia to the subtropics. The optimum sowing period for the crop is between October and November depending upon the specific region (Krishnamurthy, 1986). Under normal sowing condition, a greater part of the vegetative phase is completed, when atmospheric temperature is relatively high, but at the time of flowering the temperature is low and later on as the crops reach up to maturity the temperature and photoperiod gradually increases (Nanda *et al.*, 1995). The duration of the vegetative phase and time of season, at which seed development determinants of the yield. The high temperature and long photoperiods are also detrimental for siliquae and seed development (Munsi and Kumar 1994).

In the North-east region of India, mustard crops, with the introduction of late maturity varieties of paddy and sowing of mustard and physiological development of crops is affected due to variation in the environmental conditions, however, for most of crops, simple models based on the temperature alone can often explain 95 per cent variability in the phenological development.

The oil collections are represented by good diversity of families, species and accessions. The biggest variability of species possesses the Brassicaceae family. Some wild species of the genera Brassicaceae are scientifically studied, because of the high quality of their oils.

Recently, an increasing interest of plant breeders in wild oil bearing species has been observed, especially in terms of their higher adaptability/stress resistance to abiotic and biotic factors in the environmental conditions. Frequently, direct use of wild oil-bearing species is connected with the obtaining of various valuable products used by tradition or as a modern partial attitude to nature. Around 50 genera and over 150 species were described from the Bulgarian flora. The most studied and used species determined on the basis of their frequency distribution in nature, as well as on their economic importance of Brassicaceae family.

The oil of mustard seeds has the different characteristics. Because these oil seeds contain the significant amounts of both, saturated and un-saturated fatty acids (Lander and Morrison, 1962). Besides these, they also contain an appreciable amount of proteins, carbohydrates, vitamins and minerals status. The oil of these seeds are used for the edible as well as industrial proposes. *Brassica* seeds contain the highest amounts of crude fibre (13.6 per cent). These values are in close agreement with the ratio as reported in Indian *Brassica*. The amount of free sugars in *Brassica* is reported to be 11.0 per cent and 0.43 per cent reducing sugars. These results are very similar to those reported for Indian seeds by Krishnamurthy *et al.* (1960). The Vitamin-C was reported to be present in *Brassica* seeds and it was observed to be 0.50 per cent mg. The result indicated that, these seeds are not only a good source of vitamin -C, but also may be considered as good source of calcium and phosphorus, so can be used as a fertilizer.

Sometimes, these oil seeds are used in the confectionary and for making margarine in the country. They can also be used in the manufacture of soaps, cosmetics, perfumes, insecticides and pharmaceutical products. The oil seeds have some importance in the treatments of cold, cough and bronchial infections, inflammation of the urinary tract, gonorrohea and dirrhoea and also used in the relief of local inflammation and ulcers (Gutkin, 1950).

The work on the effects of UV-B (280-315 nm) radiation, on the crop plants and natural vegetation has been carried out by various workers. Under investigation, most of studies have been carried out to evaluate the deleterious effects of UV-B exposure on the crop plants. Agricultural scientists have been responded with a series of pioneering investigation of the hazardous or deleterious impacts of the artificial and solar UV-B radiation upon plant growth and development. A great variety of physiological and morphological plant responses to UV- radiation have been subsequently demonstrated in the past years.

Earlier studies have focussed on the independent effects of UV-B radiation on plants (Demchik and Day 1996; Smith *et al.*, 2000). However several abiotic factors altered or modified the response of plants to UV-B because of their interactions. The interaction between UV-B radiation and other environmental stresses such as herbicides (Kulandaivelu and Annamalainathan 1991), high temperature (Nedunchenzhian et *al*. 1995), freezing (Dunning *et al.*, 1994), drought (Petropoulou *et al*. 1995) and mineral deficiency (Murali and Teramura 1985) have reduced the UV-B damage to plants as compared to non-treated control plants. The UV-B damage to plants has been correlated to the nitrogen status of the plants (Hunt and McNeil 1998), plant growth increases at a high nutrient availability (Levizou and Manetas, 2001).

Thirteen crop plants were subjected to an enhanced UV-B (280-315 nm) region that simulated a 0.19 atm. cm. Ozone level, solar angle 55° (approximately 50 per cent ozone depletion at 30° N latitude), to determine their susceptibility to photosynthetic impairment. As evidenced by net CO_2 uptake and dry weight measurements, these plants exhibited a wide range of responses to enhanced levels of UV-B radiation. These species were classified either as "sensitive" *viz.* pea (*Pisum sativum*), mustard (*Brassica*), soybean (*Glycine max*) and oat (*Avena sativa*) "moderately sensitive" tomato (*Lycopersicum esculentum*), sorghum (*Sorghum bicolar*), rye (*Secale cereale*), rice (*Oryza sativa*), *tolerant corn* (*Zea mays*), pearl millet (*Pennisuetum americium*), pea nut (*Arachis hypogaea*) in respect to their sensitivity to UV-B radiation (T.K. Van *et al.*, 1976).

All the effects of elevated UV-B considered in the contest of other factors such as water stress, increase troposphere air pollution and temperature. The effects of UV-B on plants conducted mostly under growth chamber and green house conditions (Krupa, 1989). The growth of the many crop plants species such as soybean, winter wheat, rice, sorghum, cotton, corn and winter mustard etc. species has deleterious impacts by enhanced level of the UV-B radiation. The response of UV-B varies in different plant species, some are very sensitive and some are least sensitive.

The effects of UV-B radiation on the plant growth and productivity varies seasonally and is affected by microclimate and soil fertility. For instance, soybeans are less susceptible to UV-B radiation under water stress or mineral deficiency, but sensitivity increases under low levels of visible radiation (Teramura, 1983). Continued studies over many growing seasons are crucial in UV-B impact assessment of agricultural productivity. The general principal in the study to determine the deleterious effects of UV-B on crop plants that involves the use of a UV-source (lamp) coupled with different types of filters to exclude bands of UV wavelength. The intensity of UV is varied by changing the height between the lamp source and the plant canopy, because different biological processes exhibit different degrees of sensitivity to different wavelength of UV-B radiation.

The response of plants to enhanced UV-B radiation has recently been the subject of many studies was observed by Gonzalez *el al.* (1998); yang *et al*; (2005). Plant species that are sensitive to UV-B radiation often exhibit changes in the morphological traits, for instance reduction in plant height, biomass accumulation and leaf area, when exposed to increased UV-B radiation.

The UV-B radiation induced a range of morphogenic effects in the plants and these includes leaf thickening, cotyledon curling, inhibition of the hypocotyl, stem and leaf elongation, axilliary branching and shifts in the root-shoot ratio (Wilson and Greenberg 1993; Jansen *et al*;1998 and Boccalandro *et al*; 2001). Some of these responses constitute a stimulation of the growth of specific tissue (axilliary branching, leaf thickening), while others reflect an inhibition of growth (diminished hypocotyl elongation).The levels of UV-B, UV-A and PAR as well as other experimental conditions, all affects the morphogenic responses, making it difficult to compare, result from different acclimation studies.

The effects of UV-B radiation on the crop plants, includes reduction in the yield and quality, alteration in species competition, photosynthetic activity, susceptibility to disease and changes in the plant structure and pigmentation (Tevini and Teramura *et al.*, 1989), some 300 species and cultivars were tested appear to be susceptible to damage from the increased UV-B radiation. The physiological properties such as damage to Photosystem-II (Heinrich *et al.*, 1999); and reduction in the photosynthetic rate (Bassman *et al.*, 2003., Feng *et al.*, 2003., Sullivan *et al.*, 2003) and increased the activity of antioxidant enzymes such as catalase and ascorbate peroxides (APX) observed by Kim *et al.* (1996), Mazza *et al.* (1999). The UV-B radiation damage to genetic material has also been reported by Taylor *et al.* (1996), Hidema *et al.* (2000).

Ultraviolet-B radiation (280-315nm) is highly energetic and can cause damage to a wide range of cellular components, such as DNA, amino acids and lipids. Plants because of their sessile nature are potentially very susceptible to UV-B exposure and the increase in UV-B as a result of ozone depletion could have severe consequences. Over the last few decades there has been a substantial amount of studies on the effect of UV-B exposure were showed in the plants. These studies show a large number of responses including changes in growth and development, increase in protective pigment biosynthesis, effects on photosynthesis, DNA damage and cellular changes induced by UV-B exposure such as changes in gene expression. These responses are also variable, both between species and even within varieties of the some species.

The effects of UV-B radiations on the plants include reduced biomass allocation, altered biomass allocation and increased flavonoid content (Schumacher *et al.* 1997; Lavola 1998; Hofmann *et al.* 2001; Kolb *et al.* 2001). The growth responses of the plant species to ambient UV-B radiation can be positive or negative (Lakso and Huttuncen *et al.*, 1998). These differences in UV-B responses may be associated with the genetic variation, influence of other factors such as drought and nutrient availability, the presence of protectively features like waxy and reflective layers or a thick epidermis and the presence of UV-B absorbing flavonoids (Correia *et al.*, 2000; Yuan *et al.*, 2000; Beiza and Lois 2001). Although the effects of UV-B radiation on the symbiotic performance of legumes have been investigated by Chimphango *et al.* (2003) but such studies have focussed mainly on food grain legumes. The effects of UV-B radiation on tree and shrub legume nodulation and nitrogen (N_2) fixation.

The most of UV-B studies was investigated on the plants have been focussed on crops and annual plants (Papadopoulos *et al.* 1999; Kakani *et al.*, 2003). Although forests account for over two thirds of the global net primary productivity and agricultural land accounts for only about 11 per cent (Barnes *et al.*, 1993), there have been relatively few studies on the direct effects of UV-B radiation on the tress and forest ecosystem. The recent studies on the woody plants, which have been conducted both in green houses and in the field, have demonstrated several types of growth and physiological responses to enhanced UV-B radiation (Elena *et al.*, 2000; Sullivan *et al.*, 2003; Yang *et al.*, 2005).

Regarding impacts of ambient UV-B, two generalizations are beginning to emerge from field experiments, carried out in natural and cultivated ecosystems. First, ambient UV-B at the mid-latitudes appears to have measurable effects, reducing plant growth, particularly in case of herbaceous plants (Krizek *et al.*, 1998). The second modification of ambient UV-B levels may have the large impacts on the interaction between plants and phytophagous insects (Ballare *et al.*, 2001). Although there is variation among ecosystems, most common effects of solar UV-B is increased plant resistance to insects measured in terms of leaf area consumed (Zavala *et al.*, 2001).

The effects of UV-B radiation on water relations, leaf development and gas-exchange characteristics in pea (*Pisum sativum*) plants subjected to drought were investigated. Plants grown throughout their development under a high irradiance of UV-B radiation were compared with those grown without UV-B radiation. The UV-B radiation resulted in a decrease of adaxial stomatal conductance by approximately 65 per cent, increasing stomatal limitation of CO_2 uptake by 10 to 15 per cent. The growth in UV-B radiation resulted in large reductions of leaf area and plant biomass, which were associated with a decline in leaf cell numbers and cell divisions. UV-B radiation also inhibited epidermal cell expansion of the exposed surface of leaves. There was an interaction between UV-B radiation and drought treatments. UV-B radiation both delayed and reduced the severity of drought stress through reductions in plant water-loss rates, stomatal conductance and leaf area.

Photosynthetic rate and productivity in many plants species can be reduced by increased exposure to UV-B radiation (Teramura and Ziska, 1996). UV-B induced inhibition of CO_2 assimilation in mature leaves. A reduction in Rubisco activity has been suggested as a cause of the reduced CO_2 assimilation rate in leaves exposed to increased UV-B radiation. The prolonged exposure to elevated levels of UV-B radiation has been demonstrated to result in decreases of both Rubisco activity and content (Strid *et al.*, 1990; Jordan *et al.*, 1992; Kulandaivelu and Nedunchezhian, 1993). A primary cause of the decrease in the light-saturated rate of CO_2 assimilation induced by exposure to elevated UV-B radiation in leaves of oilseed rape has been shown to be a loss of Rubisco (Allen *et al.*, 1997), which may also be associated with the loss of activity of other Calvin cycle enzymes (Baker *et al.*, 1997). Exposure to UV-B radiation can modify the speed of stomatal opening and closing and reduce the rate of leaf transpiration (Negash, 1987; Day and Vogelmann, 1995). It is possible that UV-B induced effects on stomata could modify plant water relations and growth.

The leaves of *zea mays* were subjected to different scenarios of ultraviolet-B radiation in a Sun simulator to determine the cellular vitality at the microscopic level and the contents of carbohydrates and photosynthetic pigments. The result showed that leaf morphology and structure of maize leaves are slightly affected by UV-irradiance. At the microscopic level a number of epidermal cells, predominantly below leaf tips are affected by supplemental UV-B exposure. They partially collapsed or deformed cells wall and membranes. The carbohydrate partitioning was more significantly influenced by UV-treatments. The glucose was decreased under high

UV-B. Changes in the photosynthetic pigments were limited to a slight destructive effect of UV-B on the chlorophyll b. (Michael Barsing *et al.*, 2000).

Amongst these, there is now a substantial body of evidence, indicating significant effects of UV-B on secondary compounds of the Phenyl propanoid pathway and the enzymes responsible for their synthesis (Tattini *et al.*, 2000; Liakaura *et al.*, 2001), which act as sun-screens, providing protection against UV-B radiation. In these processes, monoterpense may also play a role in UV-protection (Bosabalidis and Skoula; 1998) and recently have there been several reports in case of the aromatic plants, particularly members of labiatae. Many of these plants have their centres of origin in the Mediterranean, North Africa or Asia and their normal environments are characterized by high levels of UV-B, significant induction of both terpenoids and phenylpropanoid have been demonstrated in the basil (Raki and Jonson *et al.*, 2001) and positive effects of UV-B have also been reported in the *Mentha* (Maffei and Scannerini; 2000).

In aromatic plants, the possible involvement in relation to the increasing production of the fresh herbs to meet market demand for such daily treatments with supplementary UV-B, lead to a substantial induction of essential oils in the developing plants of a Eugenol variety conditions. Scannerini *et al.* (2000) have also reported a small positive effects of UV-B on essential oil content in *Mentha piperita*, grown in controlled environment (*i.e.* lacking UV-B). Both of these reports indicate changes not only in the total content, but also in the quantitative (but not qualitative) composition. Interestingly, an enhancing effect of UV-B has also been obtained (over a much longer growing periods) in *Mentha spicata*, for plants growing in natural conditions in a Mediterranean climate, apparently even under these conditions, supplementary UV-B (given for entire seasons) can lead to enhanced levels of volatiles in the plants (Karou *et al.*, 1998).The concentration of those compounds is thought to reduce the epidermal transmittance and provides greater protection to the leaf mesophyll.

Increase in UV-B observing compounds upon UV-B exposure of plant growth, however have been correlated with reductions in transmittance (Cen and Bronman, 1993) such as photosystem-II inhibition in grape (*Vitis vinifera*) leaves was also decreased in parallel with increased epidermal screening (Kolb *et al.*, 2001). Even roots of plants, whose shoots are exposed to elevated UV-B radiation, can be affected as indicated by root interactions with microorganisms. For example the nature of microorganism assemblages that were associated with root of sugar maple tress (*Acer saccharum*) was altered by exposure of tree shoots to elevated UV-B radiation.

This was obviously a systematic effect of UV-B expressed in the root of host plant. The timing of life phases of plants is a combination of response to environmental factors and the genetic constitution of plants. Increased UV-B has been shown to advance or delay (depending on species), the timing of flowering in the plants. In the higher plants, secondary compounds, such as lignin are important as structural materials. Those are related to phenolic compounds and may change in composition with elevated UV-B radiation (Gehrke *et al.*, 1995). If the ratio of lignin to cellulose in plant tissues changes, it can alter the rate of decomposition.

It has been well established that the plant growth regulators (PGRs), influence the growth and development of plants. These chemical substances are able to coordinate growth among different plant parts or different physiological and biochemical processes. These are chemical substances known as hormone or phytohormone. The main naturally occurring plant growth hormones *viz.* IAA, Kn and GA_3 are able to control many of the physiological processes that involved in the plant development. Plant growth regulators have been tried to improve growth and ultimately yield (Ram *et al.*, 1973; Patil *et al.*, 1987 and Kumar *et al.*, 1996), tried various growth regulators to obtain better yield of good quality heads in cabbage (*Brassica* tolerance var) and obtained encouraging result. The maturity of the vegetable crops is hastened, due to the application of plant growth regulators (Buckovac and Wittwer, 1957; Chnonkar and Jha, 1963).

The phytohormones play a regulatory role in the imposition of seed dormancy, in the release of seed from dormancy, in the reserve mobilizations during seed germination and in the subsequent development, cytokines role in the seed germination (Khan and Tao, 1978). The overall plant growth was improved by the plant growth regulator treatments as compared to the UV-B treatment only. These treatments significantly, increased all plant growth parameters. The increased vegetative growth of plants nourished and developed in a better manner, than the nourishment to UV-B alone. IAA, Kn and $GA_{3,}$ which are most important growth regulators and has a profound effect on the crop production, through increase in the stem length, leaf area, flower induction, yield and weight and size of crops.

There are numerous studies on the effect of growth hormones on plants (Jawanda *et al.*, 1979; Mishra *et al.*, 1986 and Reis *et al.*, 2000). Some of these studies have shown physiological and growth traits and many have found promotion in these traits in response to increased growth hormone. The impact of growth regulators on various physiological parameters have been worked out by various workers. Mahmud (1983) evaluated the effect of various growth regulators on growth, development and yield of various varieties of oil-seed crops. The treatments of different growth substances have given remarkably encouraging results in promoting seed germination in tomato, bottle gourd, radish, lettuce, watermelon, brinjal, carrot and a number of other vegetables (Swaminathan, 1987). Vaithialingam and Rao (1973) evaluated the effects of pre-soaking of seeds of groundnut on growth and development.

Application of IAA restored the radial growth of the internode without affecting cell elongation or transverse cell divisions. Auxins in particular, are implicated in development processes like elongation growth, photo and gravitropism, apical dominance and lateral root initiation (Normanly, 1997). UV-B radiation can impact on auxin metabolism. Indole-3-acetic acid (IAA) is transported basipetally from shoot apex and young leaves to lower tissues. The distribution of auxin (IAA) is also significantly altered in seedlings of the *Arabidopsis tt4*- line, with less IAA accumulating in the upper hypocotyl the transition zone and the lower root and more in the upper root (Murphy *et al.*, 2000). The effect of IAA on spore germination in *Funaria* was reported by Chaudhary and Vyas (2002). In intraspecific crosses of *Brassica*, enhanced siliqua and seed production was reported by Sharma *et al.*,(1997) due to exogenous application of NAA, GA_3 and Kinetin.

The overall effects on elongation were the net result of these two opposite effects. Both IAA and GA are produced by the apical bud (Jones and Phillips 1996; White *et al.*, 1975). Thus, it is conceivable that internode elongation is modulated by the apex by way of a balance between the primitive and inhibitory effects of GA and IAA on cell division, on one hand and by their synergistic effects on cell elongation on the other. The effects of GA and IAA in dwarf bean differ from those reported for dwarf pea. In dwarf pea, both hormones enhanced internode elongation, when applied separately, but when applied simultaneously, IAA did not alter and even promoted the elongation elicited by GA (Arney and Mancinelli, 1969).

Cytokinins have also been reported to release dormancy and enhanced germination. The endogenous cytokinins (Kn) would appear to be key factors in the initiation of the radicle growth. The external application of cytokinins and gibberellins has been shown to substitute for the physiological influence of roots on the growth of de-rooted oat (Jordan and Skoog, 1971) and in the soybean seedlings (Holm and Key, 1969).

Kinetin used as seed treatment or foliar spray individually or in combination, increased the seed yield by 26 per cent, while foliar spray increased it by 43.6 per cent over control. However, maximum favorable effects of kinetin (Kn) were obtained with combined application of seed treatment plus foliar spray, which increased seed yield by 68.6 per cent over untreated plants (Shang *et al.*, 2000). Kinetin (Kn) also favorably affected two important plant processes *viz.* photosynthesis and nitrogen metabolism. Net photosynthetic rate and nitrate reductase activity significantly increased in plant treated with kinetin. The significant increase in content of the total chlorophyll content with kinetin application as also reported by (Khalil and Mandurahi, 1989) may also be responsible for the increase in photosynthesis (Gzik *et al.*, 1987). The concentrations of starch, soluble protein and free amino acids were maximum, when kinetin (Kn) was applied both as seed treatment and foliar spray. This could be due to kinetin mediated increase in photosynthetic and nitrate assimilation activity, besides decrease in protease activity and immobilization of nutrients and metabolites from Kn treated tissues as observed by Kumari and Bharti, (1992). The anti-oxidative activities of SOD and enzymes increase progressively with flower development, but declined during the advance stage of petal senescence of gladiolus (Hossaian *et al.*, 2006) and in rose (Kumar *et al.*, 2007).

Few attempts have been made to investigate the role of endogenous cytokinins (Kn) on the lateral bud-growth and apical dominance. The most of the hypotheses concerning the possible involvement of these hormones in apical dominance were formulated after conducting experiments with synthetic compounds. This is probably the main reason for the views that are presently held regarding the role of cytokinins in the control of the apical dominance and growth of lateral bud. As in rose, the levels of those cytokinins, which are frequently regarded as active forms of the hormones, increased with the development of shoot apex. Auxins and gibberellins (White *et al.*, 1975; Jones and Phillips, 1966) are produced by the apical bud as well as by leaves and both hormones have been suggested to be growth factors, which might regulate growth (Thimann, 1997). The central role of gibberellins (GA_3) as promoters of mitotic activity in *Caulescent* plant is well documented.

The Kinetin mediated increase in seed yield under water stress has also been reported for wheat. Comparatively, more height of Kinetin treated plants also indicates the beneficial effects in general on plant growth. The Kn application was associated with a high Harvest Index (HI), thereby, indicating partitioning of more photosynthates towards seeds. Significantly higher seed yield in Kn treated plants also led to higher water use efficiency (WUE) in spite of lower or comparable water use in control and water treated plants (Blackman and Davies, 1985). In a series of experiments,Mok (1994) observed that a large number of plant developmental processes have been found to be influenced by the cytokinin effect on cell expansion, inhibition of leaf senescence, chloroplast development, root and shoot branching. Nagel *et al.* (2001) have evaluated that cytokinin application plays a significant role in the flower production and exerted a positive effect on the yield of soybean thus increasing the total seed production. Skoog and Miller (1957) evaluated that the ratio of cytokinin to auxin in nutrient media profoundly influences the morphogenesis of roots and shoots. The characteristics property of the cytokinin to effect cell division, assimilate transport and protein synthesis in the growth of fruit was shown by Antognozzi *et al.* (1996). A synthetic cytokinin has been found to be effective in inducing parthenocarpy and in enhancing in fruit enlargements by stimulating cell division in many kinds of fruits such as apple, kiwifruit, loquat and watermelon (Hayada *et al.*, 1958 and Yu., 1999).

Purkayastha *et al.* (2008) reported that through *in vitro* culture of nodal explants, 15 days old aseptic seedling has been developed and indentified that among the various cytokinin proved to be the most effective. By the application of kinetin, a large number of shoot buds were reported to arise from the nodal segments of *Pluchea lanceolata* was identified by Arya and Patni (2007). Influence of cytokinins on the embryo formation of *Oncidium* was studied by Fan Wu *et al.* (2004) and observed that kinetin and Zeatin were found to be more effective than other auxin and cytokinin treatements to induce somatic embryogenesis from root- derived callus.

The GA did not affect cell expansion but strongly promoted transverse cell divisions and consequently increased the internode length. The similar effects of GA and IAA on the internode elongation have been reported by Lockhart (1964) and Phillips (1972). In the dwarf pea (Brian and Hemming, 1958) and cucumber (Sandhu and Kasper Baver, 1974) were observed that IAA and GA, both hormones are promoted internode elongation in decapitated plants. Therefore, the possibility that auxin have a role in internode elongation cannot be excluded. The dwarf bean plants confirm the central importance of GA in inducing the mitotic activity necessary for elongation, but they also indicate a probable dual role for auxin in the regulation of internode elongation. When IAA was applied simultaneously with GA to the decapitated internode, two opposite effects on the longitudinal growth components that determine the final internode length were observed an enhancement in cell elongation and an auxin- concentration-dependent inhibition of the transverse cell divisions. An inhibitory effect of IAA on GA-promoted elongation was also observed in non-decapitated dwarf bean plants (Shein and Jackson 1971; Vallo and Schwabe, 1978) and in internode sections of *Avena* as reported by Rapoport *et al.* (1978).

Exogenous level of gibberellins partially substituted for the cold requirement for flower stalk elongation, also implicating the involvement of plant hormone in the processes (Hanks, 1982). During rapid flower stalk growth in tulips elongation occurs mainly due to cell expansion (Gilford and Rees, 1973). The cell expansion requires readily available hexose substrates used for increased metabolism and biosynthesis of new material such as cell wall. The gibberellins have been observed to influence the carbohydrate status in the many plant species (Canomedrano *et al.*, 1997; Yim *et al.*, 1997). In elongated tissues, common response to exogenous gibberellins is an increase in acid invertase activity (Kaufmann *et al.* 1968; Miyamoto *et al.* 1993; Wu *et al.* 1993). The investigations of gibbreellins mediated changes in carbohydrate metabolism will provide useful information for better understanding the flower stalk elongation processes in tulips.

The enhanced number of open flowers per spike induced by GA_3 and sucrose treatment can also be attributed to higher petal sugar status (reducing and non-reducing sugars) and higher solution uptake. High petal sugar status and water balance in flowers is suggested to improve bud opening (Halevy and Mayak, 1981). Waithaka *et al.* (2001) was reported that the opening of gladiolus florets was accompanied by an increase in fresh-weight, dry weight and carbohydrate concentration in the preianth. The similar observation suggested by Han (2001) that was improve the bud opening in lilies, when treated with GA_3 and sucrose. Gibberellins has the characteristics property to improve the yield, plant height, flower of *chrysanthemum* as shown by Mohariya *et al.* (2003). Pharis and king (1985) observed that gibberellins (GA) play a major role in the development of fruit set. Morever, Yuda *et al.* (1984); Zhang *et al.* (2007) were also reported fruit expansion in Japanese pear. Singh *et al.* (2008) studied the effects of post-harvest application of GA_3 and BA with sucrose, on the factors influencing petal membrane stability and vase life of Gladioli.

Pearce *et al.* (2004) evaluated that the application of various growth substances *viz.* GA, IAA etc. shoot length was found to be increased in a hybrid family of poplar. The plant height was increased by GA, while branch number per plant was increased by all growth regulators and was greatest with 500 ppm IAA. Seed yield was highest with 250 or 500 ppm IAA (Stontakey *et al.*, 1991). The interaction of plant growth regulators has significant effect on shoot morphogenesis as reported by Baraldi *et al.*,(1988). Zhang *et al.* (2005) shows that various PGRs such as GA, IAA etc. play a significant role in the regulation of shoot growth and tuber formation in potato the certain evidences that the IAA promote GA biosynthesis in Barley and the role of developing inflorescence is discussed in relation to this GA interaction evaluated by Wolbang *et al.*,(2004).

During field study, it has been demonstrated by Yadav *et al.*,(2005) that the growth promoters significantly improved growth and yield of rice. Gonge *et al.* (2005) reported that due to the treatment of growth regulators production and quality characters like seed yield and germination percentage of *Okra* seeds were significantly influenced. The similar findings were also reported by Sajjan *et al.* (2003). A lot of work has been done on the effects of growth substances on the different parameters of plant growth and development by Bahuguna *et al.* (1988). Shah and Samiullah (2006) studied the effect of plant growth regulators on growth and yield of black cumin and observed that these substances were found to be more effective in promoting shoot length, dry weight, leaf number and seed yield.

All the plant growth regulators (PGRs) are probably involved at some or other stage of leaf senescence and may interact in its control, but attention has been given to the cytokinins, because of their dramatic effects, when applied exogenously (Thimann, 1980). Chinball (1954) was the first to be postulated that, some hormone like factors, produced in the roots may be involved in control of leaf senescence now called cytokinins. Variations in irradiance, light quality photoperiod and radiant exposure, produced changes in the production of endogenous growth factors. The many of these factors are necessary for the success of plant life (Moe and Anderson, 1987).

Aims and Objective

In the present research work, mitigatory impacts of plant Growth Regulators *viz.* IAA, Kn and GA_3 has been aimed to study on UV-B exposed mustard crops in lab and field study. Though, a number of references are available regarding deleterious or sometimes promotory effects of UV-B on various ecophysiological parameters as evidenced from A. Kumar (1992) and Neeta Bhatt *et al.* (2004). But no studies are available to assess the synergistic impact of UV-B and plant Growth hormones.

Therefore, present study was being proposed to evaluate the individual effects of UV-B exposure in the combination of some plant growth regulators such as IAA, Kn and GA_3 concentrations to counteract the deleterious effects of UV-B radiation with the following major objectives:

1. Seed germination and seedling growth.
2. Plant growth and development.
3. Biomass and productivity.
   - ☆ Compartmental partitioning of herbage.
   - ☆ Economic yield of the crops
4. Physiological and biochemical aspects.
   - ☆ Chlorophyll composition
   - ☆ Anthocyanin pigment
   - ☆ Enzymes: Protease and Peroxidase.

Chapter 2
Materials and Methods

Description of Study Area

Field experiments were conducted during growing seasons of 2008 to 2009 at the field of R.C.U. Government Post-Graduate College, Uttarkashi. Geographically, the District Uttarkashi is located between the Central Himalaya at 38° 28' to 31° 28' N latitude and 77° 49' to 79° 25' E longitude at an altitude of 1140 m. above mean sea level. It covers an area of 8016 km. The two major rivers of India such as Bhagirathi called Ganga and Yamuna have their origin in the snow covered peaks *viz.* Gaumukh and Banderpunch respectively, in Uttarkashi. These peaks and rivers are of high reverence of people all over India and are holy centers of pilgrimage since time immemorial. The sustenance of these mountain ranges depends upon the surroundings forest (Awasthi *et al.*, 2001). The study site was located at field main campus of P.G. College Uttarkashi. Five plots measuring 1x1 m. each were fenced by barbed wire to avoid any biotic interference. Certified seeds of two varieties of mustard crops of *Brassica compestris PT-303* (Brown Sarson) and *Brassica juncea PR-5* (Rai) were procured from Seed centre of G.B. Pant University of Agriculture and Technology Pantnagar (Uttarakhand).

Climatic Conditions of the Study Site

The climatic conditions of the Central Himalayan region are temperate in nature, which is one of the principal environmental factors, influence the interaction of biological organism and impose many of tolerance limits in the crop species, capable of surviving in the field area. Edaphic, topographic and biotic factors influence the climatic variable of light, temperature, humidity, air movement and act further and delimit the occurrence and distribution of the plant species (Johnson *et al.*, 1975). The climate of the present study area is temperate. The meteorological data of the study area are set in the Table 2.1 and Figure 2.1. The rainfall was recorded by automatic rainguage; temperature and relative humidity by automatic thermohygrograph and solar radiation were recorded by the solarimeter by author himself at the research field plots and presented in Table 2.1 and Figure 2.1. Three main seasons were clearly marked with peculiar climatic conditions.

Table 2.1: Mean maximum and minimum temperature (°C), average rainfall (mm), number of rainy days in months, relative humidity (per cent) and global solar radiation (Kcal/m²/day) of the study area during the period October 2008-2009 September.

Month	*Temperature (°C)*		*Rainfall (mm)*	*Relative Humidity (per cent)*	*Global Solar Radiation (Cal/cm²/day)*
	Maximum	*Minimum*			
October	21.3	28.6	13.0	72	748
November	17.6	25.8	37.2	68	712
December	14.5	21.7	50.3	64	630
January	7.2	17.5	10.0	73	451
February	12.4	21.4	52.0	61	569
March	18.3	28.2	22.0	54	649
April	23.2	33.7	41.0	52	551
May	25.5	36.7	100.0	47	510
June	29.5	40.5	131.0	58	545
July	26.3	37.8	392	76	472
August	27.4	34.5	274.0	82	512
September	32.2	20.5	156.0	73	625

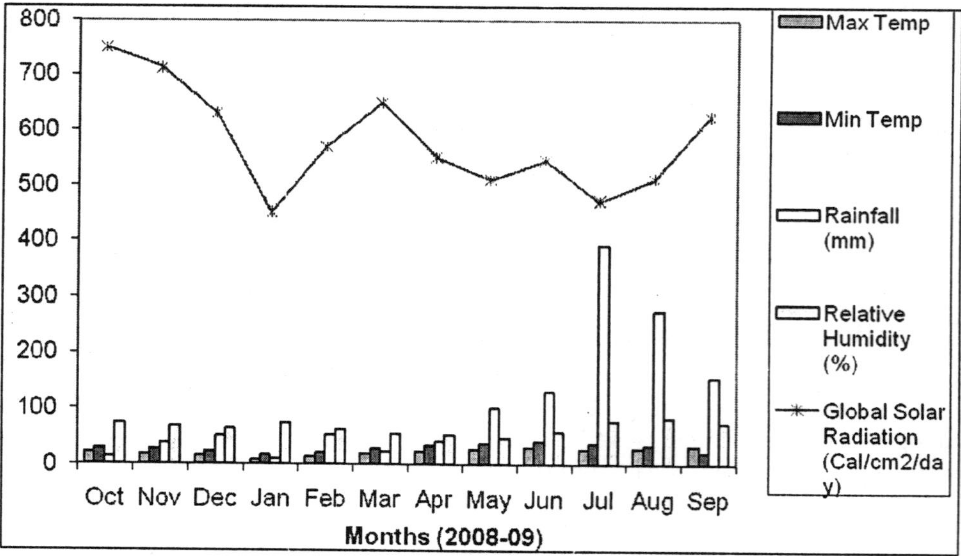

Figure 1.1: Ombrothermic Diagram of the Climatic Data of the Study Area

Edaphic Condition

Certain physical and chemical characteristics of the soil of the study plot were analyzed for the present investigation (Table 1.2).

Table 1.2: Average physical and chemical properties of the soil samples collected at three different depths of the study plot.

Soil Texture	*(Soil Depth)*		
	0-10 (cm)	*10-20 (cm)*	*20-30 (cm)*
Physical properties			
Sand (per cent)	55	47	45
Silt (per cent)	7	9	14
Clay (per cent)	39	37	34
pH	8.4	7.8	7.5
Water holding capacity (per cent)	33	44	48
Chemical properties			
Conductivity	0.08	0.05	0.11
Organic carbon, (per cent)	0.53	0.55	0.61
Total Nitrogen (N) (per cent)	0.08	0.04	0.12
Phosphorus (P) (ppm)	2.21	3.90	4.10
Potassium (K) (ppm)	207.0	210.0	215.0
Calcium (Ca) (ppm)	218.0	201.0	229.0
Sodium (Na) (ppm)	217.0	208	220.0

General Experimental Design

During laboratory studies the following sets were taken into consideration:

Control

Seeds of both varieties of mustard crops were soaked for 24 hrs. in distilled water and placed on moistened filter paper in the Petridishes.

UV-B

UV-B radiation was supplied for 3-hrs. daily by sunlamps (300W) filtered with quartz interference filters (320 nm, ORIEL, USA).

Growth Regulators

Test solution of IAA, Kn and GA_3 were prepared in three concentrations *viz.,* 10^{-7}, 10^{-5}, 10^{-6}M (in *Brassica compestris*) and 10^{-7}M (molarities) (in *Brassica juncea).*The seeds of *Brassica compestris PT 303* and *Brassica juncea PR-15* were soaked for 24 hrs. in different concentrations of growth regulators, soaked seeds were placed in paired Petridishes lined with moistened filter paper. One set of Petridish containing soaked seeds was allowed to grow without any UV-B exposure.

Growth Regulators + UV-B

In second set, one from each concentration of different growth regulators was sprayed with UV-B radiation, for 3-hrs. daily.

During field study, both varieties of mustard crops were grown in field and the plots were divided by black paper sheets into five blocks. Each field block was given treatments as follows:

Treatments of Field Plots

1. In plot-A, mustard plant species was taken as control. No treatments were given to the crop of this plot
2. Plot-B was exposed to 3-hrs. daily UV-B radiation (24.23 Jm^{-2} Z^{-1}) by Sunlamps (300W) filtered with quartz interference filters (320 nm, ORIEL, USA).
3. Plot-C was sprayed with IAA (10^{-7} M) concentration daily, along with 3-hrs supplemental UV-B radiation using the same source.
4. Plot-D was sprayed with Kn (10^{-5} M) concentration daily, along with 3-hrs. supplemental UV-B radiation by using the same source.
5. Plot-E was sprayed along with GA_3 (10^{-6} M) in *Brassica compestris PT-303* and (10^{-7} M) in *Brassica juncea PR-15* respectively along with 3-hrs. supplemental UV-B radiation, using the same source as above.

General Experimental Design may be Summarized as

In Laboratory Condition

Treat-ment	*Con-trol*	*UV-B*	*IAA*			*Kn*			*GA_3*			*IAA+UV-B*			*Kn + UV-B*			*GA_3+UV-B*		
Concentration		(3-hrs)	10^{-7}	10^{-6}	10^{-5}	10^{-7}	10^{-6}	10^{-5}	10^{-7}	10^{-6}	10^{-5}	10^{-7}	10^{-6}	10^{-5}	10^{-7}	10^{-6}	10^{-5}	10^{-7}	10^{-6}	10^{-5}

Methodology

The field for the cultivation was prepared before sowing of the seeds, as proposed by Dhasmana (1984). The pre-soaked seeds of the both varieties of mustard crops were sown in the experimental plots. The general experimental studies of different treatments were laid after complete germination of both varieties of *Brassica* crops (Kumar 1981; Ambrish 1992; N. Bhatt, 2004).

Seed Germination and Seedling Growth

For the studies of seed germination and seedling growth, uniform seeds of *Brassica compestris PT-303* (Brown Sarson) and *Brassica juncea PR-15* (Rai) were selected and surface sterilized by absolute ethyl alcohol and then 0.1 per cent $HgCl_2$ for the one minute each, thoroughly rinsed with distilled water. Total seeds of the both crops

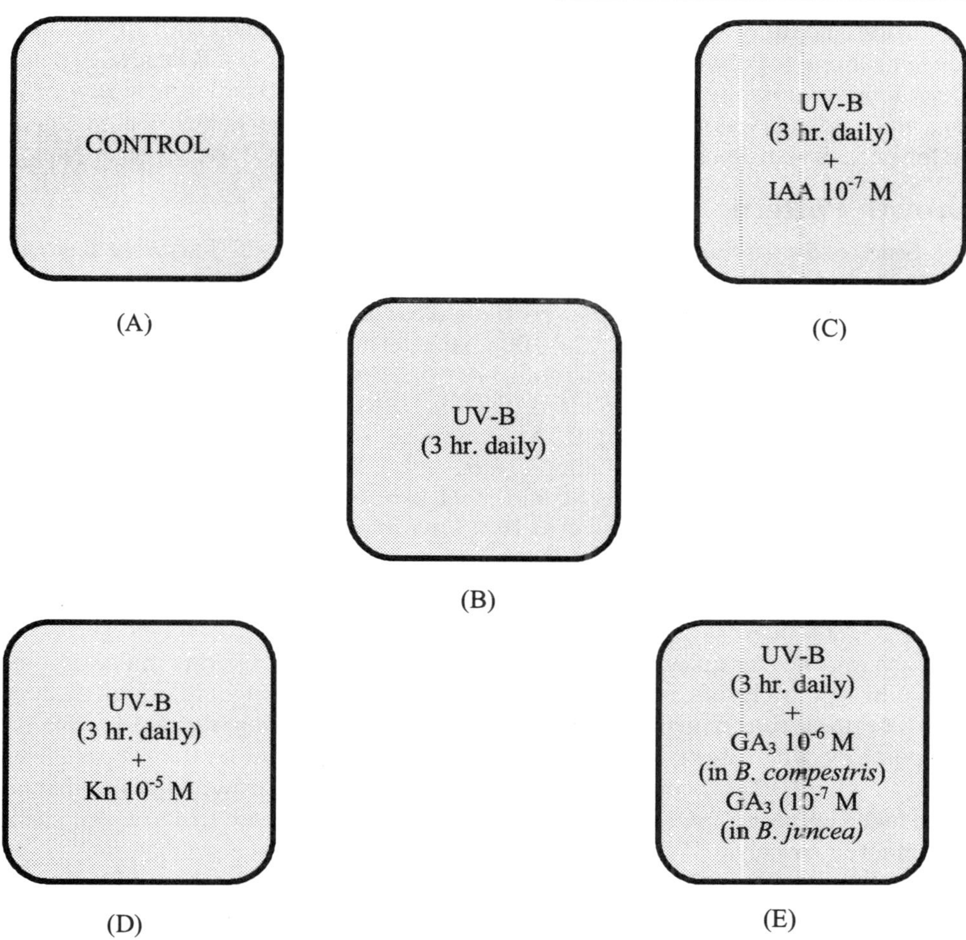

Experimental Design in the Field Plot

were divided into twenty two sets separately, *i.e.* eleven sets of each crop. Two sets of both varieties were treated as control and placed in the incubator without any treatment. Two sets of both the crops were exposed with UV-B radiation (3-hrs. daily) only.

Six sets of seeds were soaked in IAA solution of 10^{-7}, 10^{-6} and 10^{-5}M concentrations. Out of which three sets were exposed to UV-B radiation (3-hrs. daily), along with IAA 10^{-7}, 10^{-6} and 10^{-5} M concentrations.

Next six sets of seeds were soaked in the Kn solution of 10^{-7}, 10^{-6} and 10^{-5} M concentrations. Out of these, three sets were exposed to UV-B radiation (3-hrs. daily), along with Kn 10^{-7}, 10^{-6} and 10^{-5} M concentrations.

Other six sets of seeds were soaked in the GA_3 solution of 10^{-7}, 10^{-6} and 10^{-5} M. Out of these; three sets were exposed to UV-B irradiance along with GA_3 10^{-7}, 10^{-6} and 10^{-5} concentrations as given above treatments.

In the laboratory studies the all sets of Petridishes were supplied with appropriate concentrations of growth regulators daily, except control and UV-B irradiated sets. These sets were supplied with distilled water daily. Daily UV-B radiation received by seed and seedlings was about 24.23 J M^{-2} S^{-1}. The germination percentage was recorded on the basis of radicle emergence as 2 mm in length and it considered as germinated.

Growth Pattern

Seeds of *Brassica compestris PT- 303* and *Brassica juncea PR-15* (Rai) were sown in the sandy loam soil in rows placed 0.1m apart in 5 plots (A, B, C, D and E) of 1 x 1 m each. The plot-A of both the crops were treated as control and neither sprayed with growth regulators nor exposed to UV-B radiation. The plot B, C, D and E of both the crops were treated with UV-B radiation (3-hrs. daily) supplied by sunlamps (300W), filtered with quartz interference filters (ORIEL, USA). Plot C of both the crops were sprayed with IAA solution of (10^{-7} M) concentration daily, through hand spray machine, along with UV-B radiation (3-hrs. daily). Plot D of both the crops were sprayed with Kn solution of (10^{-5}M) and plot E of both the crops were sprayed with two different solutions of GA_3 such as 10^{-6}M in (*Brassica compestrisPT-303*) &10^{-7}M concentration in (*Brassica junceaPR-15*), along with UV-B radiation. The above treatments were carried out daily as described in both the crops up to maturity (harvesting period).

The samples for growth analysis were taken regularly at 15th day interval from each plot separately, after the seedling emergence (two leaf stage) till maturity. For each plot study, fifteen phenotypically identical plants were taken from field carefully to laboratory, where these were washed by running water to remove the soil particles, using a mesh of 0.32 nm pore size and tap water current. The growth measurements were taken on parts of per plant basis for stem, leaf, root, flower, fruits etc. for each treatment separately.

The mean values of 15 plants of each sample plot were calculated, results represented with ±S.D.

Biomass

Dry Matter

For the estimation of biomass and productivity "Short Term Harvest Method" (Odum, 1960) has been applied in the present study. The samples were collected on the basis of morphological similarities and standing crop was estimated on per plant basis. Fifteen plants were selected and harvested from each plot at regular interval of 15 days. The different plant parts *viz.* leaf, stem, root, flower and fruit were separated at different stages of growth. After fresh weight measurement, the samples were dried at 80°C for 48-hr, for dry matter content for each treatment separately.

Net Productivity

The net productivity was calculated by subtracting the values of biomass from that of subsequent growth stage. Summation of positive values represented the net productivity of each component. Total sum of positive values of each plant parts gives the values for the total net productivity of a crop (Odum, 1960; Bisht, 1981).

Yield Attributing Parameters

Harvesting Index (HI) and Shelling Percentage (SP) were calculated as described by Francis *et al.* (1978). Harvesting Index shows the contribution of economically useful plant part over the total production.

$$HI = \frac{\text{Economically useful plant part i.e. Fruit (g/plant)}}{\text{Total dry matter production (g/plant)}}$$

Shelling percentage shows the actual nutritive value of seed over the fruit.

$$SP = \frac{\text{Seed biomass (g/plant)}}{\text{Total fruit biomass (g/plant)}} \times 100$$

Pigment Composition

Fresh leaves (500 mg) were homogenized with 80 per cent acetone, centrifuged at 4000 rpm for 5 minutes. The filtrate was taken out and final volume was made 100 ml, using 80 per cent acetone. The optical density was read at different wavelengths *viz.* 626, 645 and 663 mm with the help of Systronics-Digital Spectrophotometer. The chlorophyll contents were estimated by the formula given by Koski and Smith (1948), which are expressed below:

Chl a,	mg /l =	12.67 A	663–2.65 A	645–0.29	626
Chl b,	mg /l =	23.60 A	645–4.23 A	663–0.33	626
Protoch,	mg /l =	29.60 A	626–3.39 A	663–6.75	645

Anthocyanins

During germination, anthocyanins were extracted by using the modified method of Mancinelli *et al.* (1975). 500 mg fresh weight of seedlings was grinded in the menthanolic HCl (80 ml methanol, 20 ml water, 1ml HCl). Homogenized tissue was transferred into glass stoppered bottle using appropriate amount of methanolic HCl, stored them overnight in the refrigerator. It was centrifuged at 4000 rpm and collected in conical flask. Final volume was made 25 ml with Methanolic HCl. Absorbance was taken at 530 nm and 660 nm, with the help of the Systronics-Digital UV-Spectrophotometer. Calculation was determined by using the formula given by Mancinelli *et al.* (1975).

As = at 530 - 1/3 A at 660

where,

As: Anthocyanins

A: Absorbance

Enzymes

Protease

Extraction

Protease enzyme was extracted in the laboratory conditions from the germinating seeds of different treatments of both the varieties of mustard crops. One gram of germinated seeds was homogenized in chilled Tris -HCl buffer, centrifuged at 5000 rpm for 5 minutes and then supernatant was used as enzyme source. The volume of supernatant was made 25 ml by adding Tris-HCl buffer. All the operations were carried out at 4 to 5° C (Sadasivam and Manickam, 1996).

Assay

Protease activity in extracted material was measured by modified method of Green and Neurath (1954). One ml of extracted material in Tris-HCl buffer, extracted earlier and preserved at 4°C, one ml of protein solution and one ml of Tris- HCl was incubated at 40°C, for 1 hrs. One ml of TCA was added to the above and kept in freezer for 3 hrs. After that, the whole solution was centrifuged to get clear supernatant. In supernatant solution, 1 ml of 1.5 N NaOH was added in a separate volumetric flask and final volume was made to 10 ml with distilled water.

One ml aliquot of the above solution was mixed with 5 ml. alkaline copper tartrate solutions and then incubated for 10 minutes at 40°C and then added one ml of Folin-phenol reagent. After 30 minutes the absorbance of the solution was read at 600 nm. A calibration curve was also prepared following the above method and utilizing standard amino acids. The released amino acids were measured through comparisons of assayed and standard curves.

Peroxidase

The extraction of crude enzyme was carried out as followed for the protease activity (Sadasivam and Manickam, 1996).

Assay

One ml aliquot of enzyme extract prepared earlier and preserved at 4° A was mixed with 7 ml distilled water, 2 ml benzidine solution, 2 ml of 6 per cent H_2O_2. The optical density of the solution was measured after 1 minute by spectrophotometer using quartz cuvettes at 610 nm. The activity measured was expressed in ".O.D. (difference) (Mahely and Chance, 1967).

Chapter 3

Deleterious Impacts of UV-B Irradiance Individually and in Combination with Certain Plant Growth Regulations (PGR_S) on Seed Germination and Seedling Growth Patterns

In the present investigation, studies for the seed germination and survival percentage under influence of UV-B individually and combined treatment of UV-B exposure and growth regulators *viz.* IAA (10^{-7} M), Kn (10^{-5} M) and GA_3 (10^{-6} M) concentrations in the *Brassica compestris PT-303* and GA_3 (10^{-7}M) in the *Brassica juncea PR-15* were observed during laboratory conditions. The 22 sets of all plant growth regulators in triplicate were prepared as discussed in materials and methods. The seed germination and survival percentage were recorded for different concentrations of plant growth regulators (PGR_S), alone and along with UV-B exposure to assess the appropriate concentrations of these growth regulators, which can be applied for the further studies and data are presented in Tables 3.1 and 3.2.

The result presented in Table 3.1 in control condition indicate that seed germination and survival percentage of *Brassica compestris PT-303* were recorded as ca. 80 per cent and 69 per cent respectively. When any one of these growth regulator was given to seedlings, the maximum germination and survival percentage were amounted as ca. 81 per cent and 72 per cent for IAA (10^{-7} M) concentration, 83 per cent and 80 per cent for Kn (10^{-5} M) concentration and 80 per cent and 76 per cent for GA_3 (10^{-6}M) concentration respectively. When the germinating seedling were treated with UV-B radiation only (3 hrs. daily), the maximum results were recorded as 78 per cent and 66 per cent for UV-B + IAA (10^{-7}M); 60 per cent and 56 per cent for UV-B + Kn (10^{-5}M); 84 per cent and 71 per cent for UV-B + GA_3 (10^{-6}M) concentration respectively. When germinating seedlings were subjected to UV-B exposure only, the germination and survival percentage was reduced and recorded as ca. 65 per cent and 59 per cent respectively.

Table 3.1: Seed germination and survival percentage of *Brassica compestris* (Brown Sarson) irradiated by UV-B radiation, individually and along with different concentrations of IAA, Kn and GA_3 (PGR_s), after 15 days of sowing.

Treatments	*Germination (per cent)*	*Survival (per cent)*
Control	80	69
UV- B only	65	59
IAA (10^{-5}) M	78	76
(10^{-6}) M	76	69
(10^{-7}) M	81	72
Kn (10^{-5}) M	83	80
(10^{-6}) M	70	67
(10^{-7}) M	72	68
GA_3 (10^{-5}) M	70	65
(10^{-6}) M	80	76
(10^{-7}) M	78	67
IAA (10^{-5} M + UV-B)	71	59
(10^{-6} M+UV-B)	77	62
(10^{-7} M +UV-B)	78	66
Kn (10^{-5} M + UV-B)	60	56
(10^{-6} M +UV-B)	59	55
(10^{-7} M +UV-B)	58	52
GA_3(10^{-5} M +UV-B)	81	70
(10^{-6} M +UVB)	84	71
(10^{-7} M +UVB)	75	68

An observation of data presented in Table 3.2 indicates that germination and survival percentage of *Brassica juncea PR-15* were 78 per cent and 65 per cent respectively, during control. When plant growth regulators (PGR_S) were supplied to germinating seedlings, the maximum germination and survival percentage were amounted to ca. 79 per cent and 76 per cent for IAA (10^{-7}M); 78 per cent and 75 per cent for Kn (10^{-5} M); 79 per cent and 74 per cent for GA_3 (10^{-7} M) concentrations respectively. When the germinating seedling were subjected to UV-B radiation only, the germination and survival percentage were recorded to ca. 61 per cent and 55 per cent respectively. When germinating seedlings were exposed to UV-B (3 hrs. daily), along with these effective concentrations of growth regulators such as IAA, Kn and GA_3 concentrations, the maximum values of seed germination and survival percentage were noted for UV-B + IAA 69 per cent and 63 per cent; for UV-B + Kn 65 per cent and 61 per cent and for UV-B + GA_3 concentration reported as ca. 66 per cent and 62 per cent respectively.

Table 3.2: Seed germination and survival percentage of *Brassica juncea PR-15* (Rai) as affected by UV-B exposure, individually and along with different concentrations of IAA, Kn and GA_3 (PGRs), after 15 days of sowing.

Treatments	*Germination (per cent)*	*Survival (per cent)*
Control	78	65
UV-B only	61	55
IAA (10^{-5}) M	76	73
(10^{-6}) M	77	75
(10^{-7}) M	79	76
Kn (10^{-5}) M	78	75
(10^{-6}) M	75	73
(10^{-7}) M	69	66
GA_3 ($10^{-5)}$ M	69	65
(10^{-6}) M	77	70
(10^{-7}) M	79	74
IAA (10^{-5} M + UV-B)	60	56
(10^{-6} M +UV-B)	68	59
(10^{-7} M + UV-B)	69	63
Kn (10^{-5} M + UV-B)	65	61
(10^{-6} M +UV-B)	59	57
(10^{-7} M +UV-B)	60	57
GA_3 (10^{-5}) + UV-B)	60	57
(10^{-6} M +UV-B)	63	59
(10^{-7} M +UV-B)	66	61

When germinating seedlings were subjected to these PGRs, such as IAA, Kn and GA_3 concentrations, the values of germination and survival percentage were identified maximum in IAA (10^{-7} M), Kn (10^{-5} M) and GA_3 (10^{-6} M) in *Brassica compestris PT-303* and 10^{-7} M concentration in *Brassica juncea PR-15* respectively. When combined treatment of PGRs and UV-B was supplied to germinating seedlings, the same concentration of these PGRs were reported effectively as compared to individual UV-B treatment, hence for the further studies IAA (10^{-7} M), Kn (10^{-5} M) and GA_3 (10^{-6} M) in *Brassica compestris PT-303* and (10^{-7} M) concentration in *Brassica juncea PR-15* with UV-B radiation (3 hrs. daily) were applied for the treatment during field studies.

The effects of UV-B radiation (3 hrs. daily), individually and in combination with different concentrations of plant growth regulators on the germinating seedlings growth of *Brassica compestris PT-303* and *Brassica juncea PR-15 viz.* epicotyl, hypocotyl and radicle showed pronounced effect of UV-B radiation and counteraction was recorded with all the PGRs, when supplied along with UV-B exposure. During laboratory condition, after 15 days of growth the length (cm), fresh and dry weight in the control were estimated as 8.25, 0.069, 0.038 for epicotyl; 6.85, 0.062, 0.032 for hypocotyl and 8.15, 0.065, 0.034 for the radicle respectively. When the germinating seedlings were exposed to UV-B radiation (3 hrs. daily), a reduction in length, fresh and dry weight of epicotyl was noted to be ca. 75.7 per cent, 88 per cent, 92 per cent; of hypocotyl 62 per cent, 90 per cent, 93.7 per cent; of radicle 76 per cent, 90.7 per cent, 94.1 per cent as compared to control (Table 3.3).

When the germinating seedlings were exposed to UV-B radiation along with different concentrations of plant growth regulators, the promotory effects was found out in all parameters as epicotyl, hypocotyl and radicle of the *Brassica compestris PT-303*. Out of these PGRs, it was noted in these concentrations *viz.* (10^{-7}M), (10^{-5}M) and (10^{-6}M) concentrations were showed a significant mitigation, against UV-B induced deleterious impacts on epicotyl, hypocotyl and radicle. In case of IAA+ UV-B, a significant increase was observed in length, fresh and dry weight of epicotyl as recorded to be ca. 93 per cent, 94 per cent, 97.4 per cent; of hypocotyl 69.3 per cent, 65 per cent, 94 per cent; of radicle 90 per cent, 97 per cent, 94.4 per cent respectively as compared to individual exposure of UV-B.

When germinating seedlings were treated with (Kn + UV-B), a significant increase was observed in length, fresh and dry weight of epicotyl and recorded to be ca. 87.2 per cent, 94.2 per cent, 94.7 per cent; for hypocotyl 70 per cent, 93.5 per cent, 93 per cent and for radicle 87.7 per cent, 94 per cent and 80 per cent respectively as compared to individual treatment of UV-B radiation. In GA_3 (10^{-6} M) concentration, when taken along with UV-B radiation, a significant increase was also observed in length, fresh and dry weight of epicotyl and amounted as 87.2 per cent, 98 per cent, 92 per cent; of hypocotyl 71 per cent, 95 per cent, 91 per cent and radicle 87 per cent, 98 per cent, 80 per cent respectively as compared to individual treatment of UV-B radiation (3 hrs. daily).

Data presented in Table 3.4 indicates that in case of *Brassica juncea PR-15,* the length, fresh and dry weight in the control condition were recorded for epicotyl as 8.55, 0.072, 0.044; for hypocotyl 6.86, 0.064, 0.041 and for radicle 8.17, 0.065, 0.043 respectively. When germinating seedlings were subjected to UV-B radiation only (3 hrs. daily), a reduction was recorded in length, fresh and dry weight of epicotyl and noted to be ca. 73 per cent, 86.1 per cent, 81 per cent; of hypocotyl 62.2 per cent, 87.5 per cent, 85.3 per cent; of radicle 76 per cent, 84 per cent and 83 per cent respectively, as compared to control condition. When the germinating seedlings were subjected to UV-B radiation along with IAA concentration, a promotion in growth of epicotyl was noticed to be ca. 90 per cent, 88 per cent, 90 per cent; of hypocotyl 69 per cent, 87.5 per cent, 85 per cent; of radicle 76.4 per cent, 84.6 per cent, 83.7 per cent respectively as compared to UV-B exposure only.

Table 3.3: Individual and combined effects of UV-B radiation (3 hrs. daily) with different concentrations (molar) of IAA, Kn and GA_3 on the seedling growth of *Brassica compestris PT-303* (Brown sarson) after 15 days of sowing.

Parameters	*Control*	*IAA*			*Kn*			*GA_3*		
		10^{-7}	*10^{-6}*	*10^{-5}*	*10^{-7}*	*10^{-6}*	*10^{-5}*	*10^{-7}*	*10^{-6}*	*10^{-5}*
					Epicotyl					
Length (cm)	8.25±0.589	8.10±0.658	7.30±1.31	7.35±1.27	7.25±1.37	7.36±1.26	8.15±0.62	7.26±1.45	8.15±0.62	7.27±1.22
F.W. (mg)	0.069±0.0052	0.066±0.051	0.065±0.034	0.067±0.013	0.66±0.033	0.065±0.025	0.067±0.024	0.064±0.023	0.067±0.025	0.065±0.035
D.W.(mg)	0.038±0.021	0.037±0.013	0.036±0.015	0.034±0.024	0.038±0.023	0.037±0.013	0.038±0.015	0.032±0.019	0.038±0.031	0.032±0.004
					Hypocotyl					
Length (cm)	6.85±1.31	5.35±0.70	5.25±0.88	5.35± 0.70	5.40±0.84	5.35±0.67	5.45±0.80	5.35±0.78	5.85±1.17	5.26±1.80
F.W. (mg)	0.062±0.041	0.060±0.034	0.059±0.039	0.059±0.034	0.061±0.015	0.059±0.043	0.060±0.047	0.058±0.020	0.061±0.018	0.056±0.035
D.W.(mg)	0.032±0.030	0.031±0.011	0.035±0.031	0.036±0.025	0.035±0.025	0.034±0.026	0.031±0.029	0.037±0.030	0.035±0.020	0.034±0.029
					Radicle					
Length (cm)	8.15±0.62	7.95±0.92	7.61±1.27	7.59±1.19	7.35±1.54	7.25±1.37	7.45±1.36	7.26±1.53	7.34±1.62	7.28±1.55
F.W. (mg)	0.065±0.041	0.059±0.034	0.058±0.042	0.056±0.021	0.055±0.015	0.054±0.029	0.056±0.030	0.054±0.040	0.055±0.024	0.057±0.007
D.W.(mg)	0.034±0.020	0.035±0.026	0.033±0.031	0.032±0.013	0.039±0.019	0.036±0.023	0.039±0.021	0.033±0.032	0.034±0.020	0.036±0.006

Contd...

Table 3.3–*Contd...*

Parameters	*UV-B only*	*IAA + UV-B*			*Kn + UV-B*			*GA_3 +UV-B*		
		10^{-7}	10^{-6}	10^{-5}	10^{-7}	10^{-6}	10^{-5}	10^{-7}	10^{-6}	10^{-5}
					Epicotyl					
Length (cm)	6.25±1.14	7.75±0.95	6.95±0.98	6.75±1.20	6.69±1.51	6.59±1.63	7.20±1.08	6.70±0.97	7.20±1.08	6.35±1.22
F.W. (mg)	0.061±0.021	0.065±0.021	0.064±0.003	0.064±0.004	0.066±0.043	0.065±0.050	0.065±0.052	0.064±0.013	0.068±0.015	0.067±0.023
D.W.(mg)	0.035±0.031	0.039±0.020	0.038±0.006	0.040±0.007	0.038±0.021	0.037±0.032	0.036±0.043	0.036±0.015	0.041±0.020	0.040±0.031
					Hypocotyl					
Length (cm)	4.25±0.46	4.75±0.046	4.40±0.30	4.45±0.28	4.75±0.58	4.76±0.59	4.86±0.78	4.35±0.62	4.87±0.79	4.15±0.52
F.W. (mg)	0.056±0.050	0.059±0.011	0.057±0.024	0.061±0.021	0.061±0.029	0.059±0.021	0.058±0.021	0.056±0.012	0.059±0.021	0.058±0.041
D.W.(mg)	0.030±0.025	0.034±0.021	0.033±0.029	0.036±0.025	0.031±0.013	0.035±0.025	0.030±0.013	0.033±0.010	0.035±0.026	0.034±0.025
					Radicle					
Length (cm)	6.20±1.22	7.35±1.13	6.85±1.11	6.70±0.91	6.65±1.02	6.57±1.12	7.15±1.13	6.65±1.02	7.15±0.78	6.25±1.33
F.W. (mg)	0.059±0.025	0.067±0.035	0.064±0.028	0.059±0.018	0.066±0.014	0.065±0.015	0.069±0.020	0.063±0.020	0.066±0.005	0.064±0.008
D.W.(mg)	0.032±0.015	0.036±0.025	0.038±0.018	0.039±0.017	0.045±0.016	0.043±0.018	0.042±0.031	0.036±0.019	0.042±0.002	0.039±0.009

Table 3.4: Individual and combined effects of UV-B radiation (3 hrs. daily) with different concentrations (molar) of IAA, Kn and GA_3 on the seedling growth of *Brassica juncea PR-15* (Rai) after 15 days of sowing.

Parameters	*Control*	*IAA*			*Kn*			*GA_3*		
		10^{-7}	10^{-6}	10^{-5}	10^{-7}	10^{-6}	10^{-5}	10^{-7}	10^{-6}	10^{-5}
					Epicotyl					
Length (cm)	8.55±0.28	8.20±0.58	7.32±1.12	7.35±1.16	7.26±1.19	7.35±1.13	8.16±0.85	7.25±1.20	8.17±0.84	7.26±1.19
F.W. (mg)	0.072±0.007	0.067±0.031	0.066±0.051	0.065±0.005	0.065±0.007	0.064±0.029	0.067±0.051	0.067±0.050	0.066±0.061	0.067±0.028
D.W.(mg)	0.044±0.008	0.035±0.022	0.034±0.003	0.035±0.009	0.034±0.025	0.032±0.026	0.035±0.009	0.031±0.021	0.032±0.031	0.036±0.015
					Hypocotyl					
Length (cm)	6.86±1.24	5.36±0.70	5.27±0.88	5.36±0.70	5.39±0.72	5.34±0.76	5.46±0.73	5.36±0.70	5.86±1.13	5.26±0.82
F.W. (mg)	0.064±0.023	0.061±0.039	0.060±0.029	0.058±0.025	0.059±0.008	0.058±0.025	0.061±0.010	0.062±0.029	0.059±0.005	0.055±0.018
D.W.(mg)	0.041±0.002	0.031±0.025	0.031±0.018	0.035±0.017	0.030±0.009	0.031±0.018	0.035±0.019	0.036±0.030	0.036±0.029	0.033±0.029
					Radicle					
Length (cm)	8.17±0.84	7.93±1.23	7.30±1.11	7.32±1.10	7.24±1.11	7.25±1.11	7.43±1.15	7.30±1.11	7.29±1.13	7.25±1.20
F.W. (mg)	0.065±0.009	0.059±0.015	0.058±0.031	0.055±0.019	0.056±0.031	0.053±0.019	0.065±0.022	0.064±0.025	0.056±0.031	0.058±0.031
D.W.(mg)	0.043±0.008	0.034±0.020	0.038±0.021	0.031±0.008	0.033±0.025	0.035±0.031	0.032±0.011	0.033±0.035	0.033±0.021	0.034±0.021

Contd...

Table 3.4–*Contd...*

Parameters	*UV-B only*	*IAA + UV-B*			*Kn + UV-B*			*GA_3 +UV-B*		
		10^{-7}	*10^{-6}*	*10^{-5}*	*10^{-7}*	*10^{-6}*	*10^{-5}*	*10^{-7}*	*10^{-6}*	*10^{-5}*
Epicotyl										
Length (cm)	**6.27**±1.11	**7.70**±0.96	**6.94**±0.87	**6.75**±1.00	**6.68**±0.88	**6.58**±0.95	**7.36**±1.26	**7.25**±1.11	**6.65**±1.14	**6.95**±0.81
F.W. (mg)	**0.062**±0.006	**0.064**±0.031	**0.065**±0.026	**0.063**±0.007	**0.059**±0.027	**0.063**±0.025	**0.064**±0.021	**0.061**±0.035	**0.058**±0.010	**0.059**±0.012
D.W. (mg)	**0.036**±0.007	**0.040**±0.003	**0.037**±0.008	**0.039**±0.026	**0.039**±0.008	**0.032**±0.031	**0.032**±0.019	**0.030**±0.001	**0.035**±0.025	**0.039**±0.018
Hypocotyl										
Length (cm)	**4.27**±0.63	**4.77**±0.76	**4.45**±0.61	**4.51**±0.62	**4.78**±0.73	**4.76**±0.72	**4.79**±0.73	**4.78**±0.73	**4.35**±0.86	**4.65**±0.83
F.W. (mg)	**0.056**±0.009	**0.060**±0.024	**0.058**±0.013	**0.060**±0.028	**0.057**±0.030	**0.058**±0.009	**0.059**±0.021	**0.059**±0.025	**0.056**±0.031	**0.058**±0.005
D.W.(mg)	**0.035**±0.021	**0.034**±0.031	**0.034**±0.029	**0.031**±0.013	**0.029**±0.020	**0.035**±0.021	**0.032**±0.007	**0.032**±0.024	**0.034**±0.025	**0.035**±0.015
Radicle										
Length (cm)	**6.25**±1.33	**7.35**±1.39	**6.90**±1.29	**6.70**±1.35	**6.65**±1.02	**6.55**±1.17	**7.23**±0.89	**7.20**±0.90	**6.61**±1.21	**6.92**±0.96
F.W. (mg)	**0.055**±0.025	**0.057**±0.009	**0.056**±0.008	**0.053±0.007**	**0.054**±0.004	**0.052**±0.021	**0.061**±0.008	**0.060**±0.001	**0.055**±0.024	**0.059**±0.029
D.W.(mg)	**0.036**±0.026	**0.043**±0.010	**0.039**±0.015	**0.036±0.021**	**0.035**±0.005	**0.036**±0.027	**0.032**±0.006	**0.039**±0.029	**0.035**±0.027	**0.036**±0.019

When germinating seedlings were exposed to UV-B radiation along with Kn concentration, a promotory effect was found in growth of epicotyl and recorded to be ca. 86 per cent, 88 per cent, 72.7 per cent; of hypocotyl 69.8 per cent, 92.1 per cent, 78 per cent; of radicle 88 per cent, 93 per cent, 74.4 per cent respectively, as compared to individual treatment of UV-B radiation. In case of GA_3 + UV-B, the promotory effects was observed in terms of the epicotyl and noted to be ca. 84.7 per cent, 84 per cent, 68.1 per cent; of hypocotyl ca. 69.6 per cent, 92 per cent, 78 per cent; of radicle ca. 88 per cent, 82 per cent and 90 per cent respectively, as compared to individual treatment of UV-B radiation.

Chapter 4

Effects of UV-B Irradiance Alone and Alongwith Certain Plant Growth Regulators on Growth Patterns

For further studies, plants of both varieties of mustard crops were grown in the field as described in materials and methods. The seeds of Brown Sarson and Rai were sown in the sandy loam soil in the different treatment plots in the growing seasons. Plot-A (control), plot-B (UV-B only), plot-C (UV-B + IAA10^{-7} M), plot-D (UV-B + Kn (10^{-5} M) and plot-E (UV-B + $GA_3 10^{-6}$ M) in *Brassica compestris PT-303* and (10^{-7} M) concentration in *Brassica juncea PR-15* were treated accordingly. The plants of both the varieties of mustard crops, for growth analysis were taken regularly and randomly at the 15 days interval from the seedling emergence stage till maturity. The results have been documented in Tables 4.1–4.5; Figures 4.1–4.3 for the *Brassica compestris PT-303* and Tables 4.6–4.10; Figures 4.4–4.6 for the *Brassica juncea PR-15.*

Brassica compestris PT-303

The data of the stem growth patterns as affected by the various treatments are presented in Table 4.1. In the control plot (A), the value of stem length (cm/pl), fresh weight and dry weight (g/pl) of the stem were recorded at the fifteen (15) day stage of the growth as 4.55 cm/pl, 0.079, 0.035 g/plant respectively and observed to be increased continuously up to maturity and noticed as 39.75 cm/pl, 0.787 and 0.544 g/plant respectively. When the plot (B) was exposed to UV-B radiation (3 hrs. daily) only, it was observed deleterious to stem length, fresh weight and dry weight of the

Table 4.1: Stem growth patterns of field grown *Brassica compestris PT-303* as affected by UV-B radiation (3 hrs. daily), individually and in combination of IAA, Kn and GA_3.

Treat-ments	Para-meters	Crop Age in Days								
		15	30	45	60	75	90	105	120	135
A	Length(cm)	4.55± 0.31	7.15±0.45	12.00±0.6	15.50±0.74	18.00±0.40	21.10± 1.100	24.10±1.197	27.45±0.761	39.75±3.32
	F.W. (g)	0.079±0.006	0.123±0.417	0.269±0.0049	4145±0.0057	0.519±0.040	0.715±0.041	0.724±0.099	0.728±0.010	0.787±0.041
	D.W. (g)	0.034±0.0032	0.102±0.0065	0.134±0.0069	0.315±0.0052	0.414±0.05	0.513±0.077	0.512±0.054	0.514±0.057	0.544±0.058
B	Length(cm)	3.75± 0.540	5.85±0.818	10.95±1.480	13.05±0.497	16.80±0.537	19.80±0.78	22.15±1.31	25.70±1.42	32.85±6.62
	F.W. (g)	0.059±0.0053	0.097±0.005	0.248±0.007	0.378±0.088	0.388±0.051	0.515±0.049	0.540±0.031	0.554±0.052	0.646±0.055
	D.W. (g)	0.024±0.0053	0.055±0.0050	0.104±0.0054	0.254±0.029	0.244±0.036	0.313±0.029	0.320±0.051	0.332±0.053	0.323±0.054
C	Length (cm)	3.90±0.459	6.20±0.258	11.40±1.021	14.95±0.586	17.30±0.483	20.25±0.823	23.05±0.0 55	26.95±0.057	35.80±4.014
	F.W. (g)	0.065±0.054	11.45±0.029	0.248±0.051	0.398±0.056	0.399±0.057	0.621±0.0047	0.682±0.0070	0.688±0.0052	0.768±0.0051
	D.W. (g)	0.032±0.055	0.057±0.048	0.124±0.038	0.264±0.057	0.276±0.045	0.341±0.0069	0.341±0.0058	0.354±0.0075	0.494±0.0053
D	Length (cm)	3.80±0.788	6.10±0.316	11.55±1.141	14.80±0.674	17.45±0.614	20.40±0.774	23.00±1.080	26.20±2.485	34.62±4.991
	F.W. (g)	0.065±0.0055	0.069±0.0072	0.218±0.0041	0.397±0.0053	0.621±0.0065	0.682±0.0052	0.685±0.0097	0.690±0.069	0.758±0.0039
	D.W. (g)	0.030±0.0053	0.055±0.0069	0.124±0.0010	0.265±0.0057	0.274±0.0049	0.312±0.029	00.341±0.0054	0.354±0.0039	0.524±0.0056
E	Length (cm)	3.90±0.459	6.95±0.550	11.45±1.065	15.00±0.527	17.60±0.714	20.50±0.707	23.90±2.011	26.95±0.685	36.89±0.0083
	F.W. (g)	0.065±0.0050	0.069±0.0061	0.228±0.0046	0.398±0.0068	0.482±0.057	0.651±0.0052	0.661±0.0040	0.676±0.0052	0.768±0.0046
	D.W. (g)	0.025±0.0035	0.056±0.0050	0.124±0.0046	0.264±0.0056	0.281±0.0069	0.321±0.0051	0.351±0.0051	0.362±0.0053	0.525±0.0151

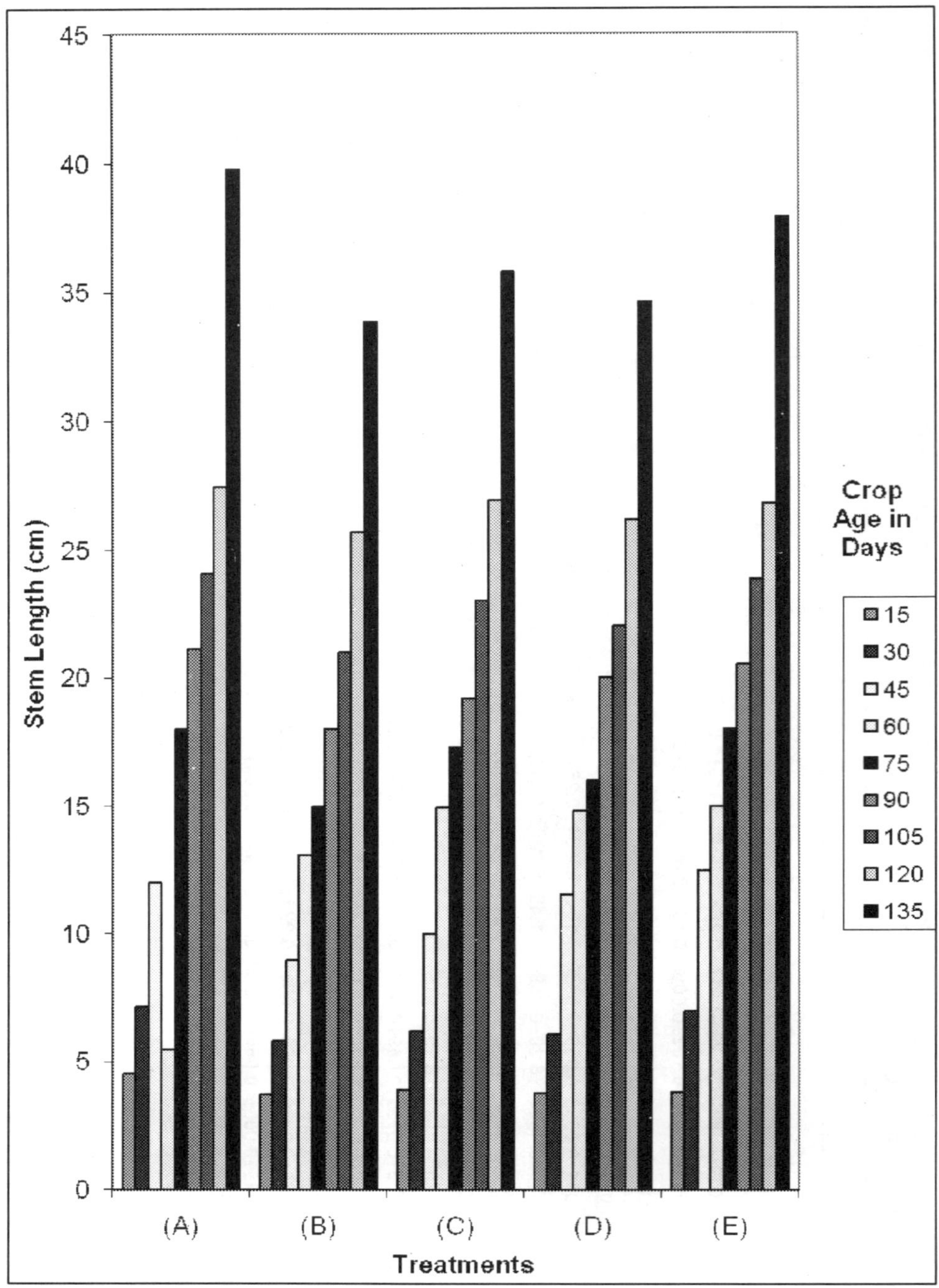

Figure 4.1: Stem growth patterns of field grown *Brassica compestris PT-303* as affected by UV-B by different treatments.

stem as compared to control. The maximum inhibition of length, fresh weight and dry weight was noticed at the 15th day stage of growth and recorded as 17 per cent, 25 per cent, 31 per cent; at the 30th day as 19 per cent, 21 per cent, 40 per cent; at the 75th day as 7 per cent, 26 per cent, 42 per cent and at the maturity as 9 per cent, 30 per cent, 40 per cent respectively as compared to control.

When UV-B treatment was given along with IAA (10^{-7}M) concentration, the promotory effect was reported and maximum mitigation of length, fresh weight and dry weight were noticed at the 60th day and recorded as 13.4 per cent, 5.3 per cent and 4.1 per cent respectively over UV-B treatment. When UV-B radiation was given along with Kn (10^{-5} M) concentration, the enhancement of length, fresh weight and dry weight were observed at the 45th day stage of growth and recorded as 5.4 per cent, 13.7 per cent, 19.1 per cent; at the 90th day as 3.3 per cent, 32.4 per cent, 24 per cent and at maturity stage as 6.2 per cent, 17.2 per cent, 61 per cent respectively as compared to individual treatment of UV-B radiation (3 hrs. daily). When UV-B exposure was also given along with GA_3 (10^{-6} M) concentration, the enhancement of length, fresh weight and dry weight were noted at the 15th day stage of growth and observed as 4 per cent, 10.1 per cent, 6.5 per cent; at the 60th day as 14.9 per cent, 5.2 per cent, 4.1 per cent; at the 90th day as 4.7 per cent, 26.4 per cent, 4.3 per cent and at the maturity observed as 5.8 per cent, 19 per cent, 62 per cent respectively as compared to individual treatment of UV-B exposure (3 hrs. daily).

The data of the leaf growth pattern as influenced by various treatments are represented in the Table 4.2. During control plot, the value of leaf area, fresh weight and dry weight were recorded at the 15th day stage of growth and amounted as 3.50 cm^2/pl, 0.115, 0.084 g/pl respectively and increased continuously up to maturity stage and measured as 35.15 cm^2/pl, 0.438 g/pl and 0.316 g/pl respectively. When UV-B exposed (3 hrs. daily) plot was studied, the all these parameters were found decreased till maturity and noticed as 2.60 cm^2/pl, 0.095 g/pl, 0.052 g/pl respectively. It was observed that the maximum inhibition of leaf area, fresh weight and dry weight were identified at the 15th day stage of growth and recorded as ca. 26 per cent, 19 per cent, 39 per cent; at the 45th day as ca. 23 per cent, 18.1 per cent, 31 per cent; at the 75th day as ca.16 per cent, 10.4 per cent, 23 per cent and at the maturity recorded as ca. 12 per cent, 6 per cent, 26 per cent respectively as compared to the control.

When UV-B exposure was given along with IAA (10^{-7} M) concentration, the promotion was observed in leaf area, fresh weight and dry weight up to maturity stage and amounted as 10 per cent, 30 per cent and 11 per cent respectively as compared to individual treatment of UV-B exposure. When UV-B exposure was given along with Kn (10^{-5}M) concentration, the promotory effects was observed in leaf area, fresh weight and dry weight up to maturity stage of growth and observed as 7.2 per cent, 24.6 per cent and 10 per cent respectively as compared to individual treatment of UV-B radiation (3 hrs. daily). When UV-B exposure was given along with GA_3 (10^{-6} M) concentration, the promotion was also observed in leaf area, fresh weight and dry weight up to maturity stage and recorded as 10 per cent, 27.5 per cent, and 23.2 per cent respectively as compared to individual treatment of UV-B radiation (3 hrs. daily).

Table 4.2: Leaf growth patterns of field grown *Brassica compestris PT-303* as affected by UV-B radiation (3 hrs. daily) individually and in combination of IAA, Kn and GA_3.

Treat-ments	*Para-meters*	*Crop Age in Days*								
		15	*30*	*45*	*60*	*75*	*90*	*105*	*120*	*135*
A	Leaf Area (cm^2)	3.50±0.824	6.97±1.060	12.09±0.747	14.97±0.573	17.67±3.017	19.75±0.925	22.26±3.070	26.86±0.44	35.15±0.85
	F.W. (g)	0.115±0.051	0.121±0.0031	0.129±0.079	0.213±0.0072	0.219±0.0060	0.313±0.0069	0.319±0.0034	0.412±0.0076	0.438±0.0053
	D.W. (g)	0.084±0.031	0.091±0.0091	0.106±0.0025	0.116±0.0071	0.156±0.0029	0.212±0.059	0.214±0.0030	0.311±0.0037	0.315±0.069
B	Leaf Area (cm^2)	2.60±1.068	5.01±0.619	9.38±0.586	12.05±0.654	14.85±0.818	16.93±0.544	20.04±4.752	24.40±0.016	31.11±4.49
	F.W. (g)	0.094±0.003	0.099±0.032	0.106±0.072	0.193±0.076	0.198±0.079	0.264±0.048	0.273±0.0091	0.316±0.0031	0.312±0.0029
	D.W. (g)	0.052±0.002	0.054±0.049	0.074±0.069	0.142±0.046	0.114±0.0073	0.142±0.043	0.142±0.038	0.213±0.029	0.234±0.0069
C	Leaf Area (cm^2)	2.72±0.760	6.02±0.893	11.35±1.49	13.22±1.356	15.03±0.592	18.07±2.901	21.10±3.60	25.00±4.907	33.46±5.085
	F.W. (g)	0.114±0.021	0.119±0.069	0.121±0.093	0.204±0.039	0.209±0.099	0.293±0.0039	0.298±0.069	0.354±0.0067	0.408±0.089
	D.W. (g)	0.064±0.035	0.069±0.060	0.082±0.029	0.152±0.0091	0.125±0.049	0.151±0.0063	0.164±0.030	0.251±0.062	0.259±0.079
D	Leaf Area (cm^2)	2.62± 0.543	6.32±0.577	10.07±1.453	13.20± 1.329	15.03± 0.544	18.03±2.908	21.70±3.43	25.05±4.095	33.36±5.085
	F.W. (g)	0.114±0.036	0.119±0.059	0.121±0.034	0.203±0.0071	0.208±0.094	0.293±0.0052	0.287±0.0032	0.346±0.0065	0.389±0.0071
	D.W. (g)	0.064±0.039	0.058±0.049	0.082±0.025	0.151±0.0039	0.126±0.093	0.151±0.023	0.165±0.0047	0.253±0.095	0.256±0.0039
E	Leaf Area (cm^2)	2.72±0.447	6.02±0.711	9.38± 0349	13.10±1.325	15.81±0.682	18.40±0.658	21.60±1.308	25.45±4.910	33.20±7.420
	F.W. (g)	0.114±0.094	0.119±0.075	0.121±0.043	0.204±0.025	0.209±0.095	0.293±0.079	0.299±0.046	0.354±0.0037	0.398±0.075
	D.W.(g)	0.064±0.083	0.059±0.038	0.081±0.026	0.152±0.0026	0.126±0.0070	0.151±0.039	0.174±0.0040	0.262±0.058	0.286±0.0072

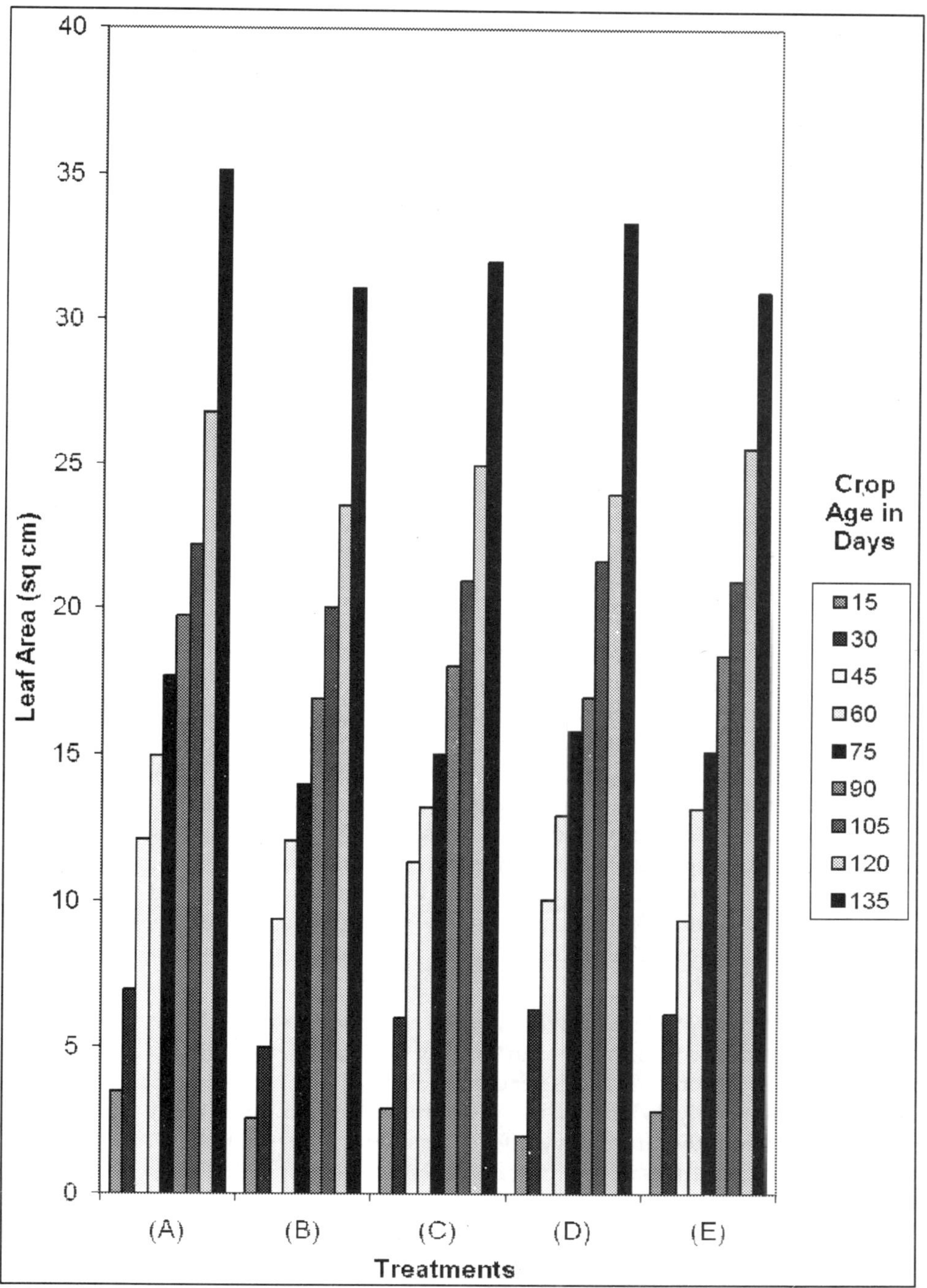

Figure 4.2: Leaf growth patterns of field grown *Brassica compestris PT-303* as affected by different treatments.

In control plot, the value of fresh weight and dry weight of root was found at the 15th day stage of growth and amounted to ca. 0.078 and 0.036 g/pl respectively and observed to be increased continuously up to maturity and amounted as ca. 1.31 and 0.82 g/pl respectively. When the plot (B) was exposed to UV-B only (3 hrs. daily), a deleterious impact was observed on fresh weight and dry weight of root as compared to control. The maximum inhibition of the fresh weight and dry weight of root was noticed at the 30th day and decreased by ca. 20.2 per cent, 53 per cent; at the 75th day decreased by ca. 34 per cent, 26 per cent and maturity decreased by ca. 14 per cent, 25 per cent respectively as compared to control condition.

When the plants were exposed to UV-B (3 hrs. daily), along with IAA (10^{-7} M) concentration, the maximum promotion was observed on the fresh weight and dry weight of roots were noted at the 120th day stage of growth and an increased by 19 per cent and 92 per cent respectively as compared to UV-B exposed plot only. When UV-B exposed plant (plot-D) was sprayed with Kn (10^{-5} M) concentration, the maximum promotory effects was recorded on the fresh and dry weight at the 120th day stage of growth and mitigated by ca. 19 per cent and 31 per cent respectively as compared to UV-B exposure (3 hrs. daily) only. When UV-B was also given along with GA_3 (10^{-6} M) concentration, the maximum promotory effects was recorded on the fresh weight and dry weight at the 135th day stage and mitigated by ca. 9.5 per cent and 16.3 per cent respectively as compared to the individual treatment of UV-B radiation (Table 4.3).

In the present investigation, flowering analysis was carried out from the 60th day stage up to maturity, in terms of fresh and dry weight (g/pl) and data were depicted in the table. In the control plot, the value of fresh and dry weight of flowers was found at the 60th day stage of growth and recorded as ca. 0.0580 and 0.0340 g/pl respectively and observed to be increased continuously up to maturity and amounted as ca. 0.0987 and 0.0730 g/pl respectively. When these plants were exposed to UV-B radiation (3 hrs. daily) only, the reductions were found in all these considered parameters as compared to control conditions. The maximum inhibition in the fresh and dry weight of flowering was observed at the 60th day and inhibited by ca. 45 per cent, 65 per cent; at the 105th day inhibited by ca. 25 per cent, 37 per cent and at the 120th day stage of growth inhibited by ca. 20 per cent, 23.3 per cent respectively as compared to the control.

When UV-B exposed plants were studied along with IAA (10^{-7}M) concentration, a mitigatory response was observed in terms of fresh and dry weight of flowers at the 60th day stage and increased by 53 per cent and 80 per cent respectively as compared to UV-B exposure only. When UV-B exposed plants were sprayed along with Kn (10^{-5}M) concentration, the promotory effects was reported from the 60th day stage of growth and increased by 53 per cent and 85 per cent respectively as compared to individual treatment of UV-B exposure. When UV-B exposed plants were studied along with GA_3 (10^{-6}M) concentration, the promotory effect was also reported at the 120th day stage of growth and increased by 13.9 per cent and 23.6 per cent respectively in terms of fresh and dry weight of flowers as compared to individual treatments of UV-B (3 hrs. daily) radiation (Table 4.4).

Table 4.3: Root growth patterns of field grown *Brassica compestris PT-303* as affected by UV-B radiation (3 hrs. daily), individually and in combination of IAA, Kn and GA_3.

Treat-ments	*Para-meters*	*Crop Age in Days*								
		15	*30*	*45*	*60*	*75*	*90*	*105*	*120*	*135*
A	F.W. (g)	0.078±0.007	0.121±0.0071	0.129±0.063	0.268±0.065	0.414±0.063	0.419±0.005	0.518±0.007	0.719±0.085	1.314±0.073
	D.W.(g)	0.035±0.003	0.090±0.003	0.109±0.073	0.234±0.035	0.310±0.044	0.314±0.063	0.413±0.008	0.512±0.039	0.822±0.075
B	F.W. (g)	0.067±0.004	0.097±0.002	0.112±0.051	0.129±0.071	0.277±0.056	0.351±0.045	0.397±0.003	0.513±0.005	1.135±0.076
	D.W.(g)	0.030±0.002	0.043±0.009	0.095±0.004	0.098±0.035	0.232±0.036	0.271±0.039	0.214±0.005	0.312±0.004	0.622±0.068
C	F.W.(g)	0.071±0.007	0.111±0.008	0.119±0.071	0.256±0.004	0.351±0.009	0.398±0.064	0.497±0.004	0.612±0.003	1.242±0.008
	D.W.(g)	0.032±0.078	0.053±0.007	0.103±0.035	0.223±0.005	0.280±0.008	0.301±0.076	0.321±0.003	0.412±0.002	0.721±0.005
D	F.W. (g)	0.072±0.035	0.112±0.069	0.118±0.021	0.256±0.006	0.362±0.003	0.398±0.009	0.496±0.002	0.612±0.001	1.246±0.005
	D.W.(g)	0.033±0.041	0.052±0.003	0.102±0.035	0.213±0.008	0.291±0.007	0.302±0.008	0.322±0.006	0.411±0.006	0.723±0.003
E	F.W.(g)	0.072±0.061	0.111±0.074	0.118±0.045	0.257±0.004	0.361±0.069	0.397±0.007	0.497±0.081	0.613±0.006	1.248±0.004
	D.W.(g)	0.032±0.053	0.053±0.052	0.103±0.039	0.223±0.003	0.291±0.096	0.298±0.079	0.331±0.095	0.420±0.007	0.724±0.007

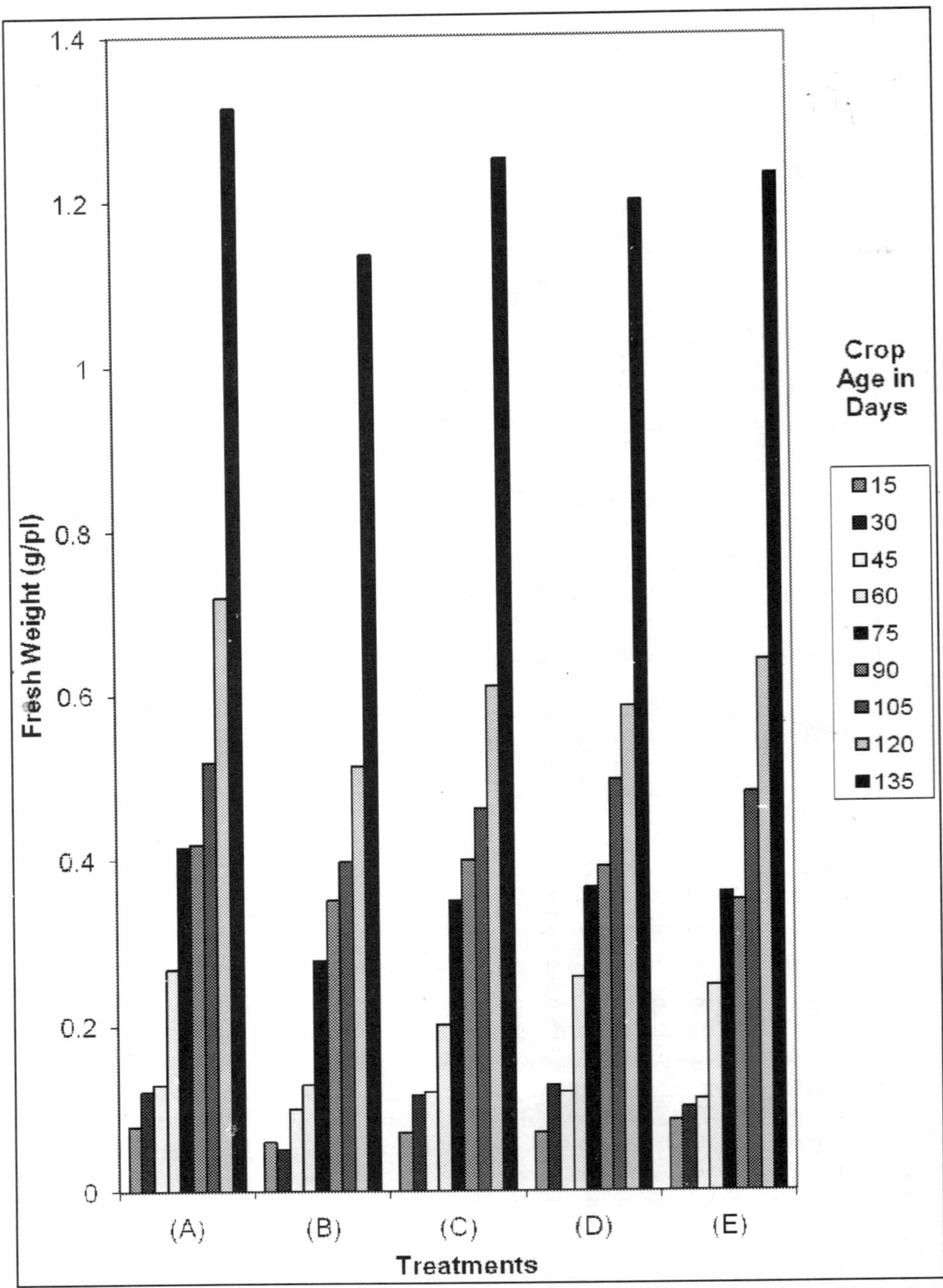

Figure 4.3: Root growth patterns of field grown *Brassica compestris PT-303* as affected by different treatments.

Table 4.4: Flower growth patterns of field grown *Brassica compestris PT-303* as affected by UV-B radiation (3 hrs. daily), individually and in combination of IAA, Kn and GA_3.

Treat-ments	*Para-meters*	*Crop Age in Days*								
		15	*30*	*45*	*60*	*75*	*90*	*105*	*120*	*135*
A	F.W. (g)	–	–	–	**0.058**±0.034	**0.071**±0.058	**0.079**±0.064	**0.086**±0.037	**0.094**±0.041	**0.116**±0.037
	D.W. (g)	–	–	–	**0.034**±0.031	**0.048**±0.047	**0.053**±0.006	**0.058**±038	**0.061**±0.035	**0.073**±0.091
B	F.W. (g)	–	–	–	**0.032**±0.006	**0.052**±0.038	**0.061**±0.038	**0.066**±0.065	**0.076**±0.081	**0.083**±0.073
	D.W. (g)	–	–	–	**0.013**±0.069	**0.036**±0.076	**0.039**±0.029	**0.044**±0.0051	**0.053**±0.071	**0.066**±0.064
C	F.W.(g)	–	–	–	**0.049**±0.047	**0.066**±0.057	**0.069**±0.031	**0.073**±0.051	**0.079**±0.069	**0.098**±0.049
	D.W. (g)	–	–	–	**0.022**±0.009	**0.029**±0.037	**0.039**±0.041	**0.052**±0.041	**0.059**±0.036	**0.069**±0.043
D	F.W. (g)	–	–	–	**0.049**±0.059	**0.065**±0.063	**0.069**±0.009	**0.068**±0.025	**0.078**±0.009	**0.098**±0.029
	D.W. (g)	–	–	–	**0.022**±0.0078	**0.044**±0.014	**0.047**±0.007	**0.043**±0.041	**0.054**±0.007	**0.056**±0.031
E	F.W. (g)	–	–	–	**0.049**±0.069	**0.066**±0.031	**0.069**±0.079	**0.068**±0.058	**0.078**±0.063	**0.097**±0.035
	D.W. (g)	–	–	–	**0.022**±0.058	**0.044**±0.009	**0.048**±0.068	**0.053**±0.032	**0.055**±0.054	**0.064**±0.031

Fruits determined the yield of any crop plant. In the present investigation, the fruiting analysis was carried out from the 75th day stage up to maturity in terms of the fresh weight and dry weight (g/pl) of fruits. The data indicated that UV-B radiation significantly inhibited the growth of fruit but showed the promotory effects, when these varieties of mustard crops was exposed to UV-B in the combination of IAA (10^{-7} M), Kn (10^{-5} M) and GA_3 (10^{-6} M) concentrations of plant growth regulators (PGRs). When these plants were exposed to UV-B (3 hrs. daily) only, a reduction was observed in fruits as compared to the control plot. The maximum inhibition in the fresh and dry weight of fruit was recorded at the 75th day and inhibited by 50 per cent, 58 per cent; at the 105th day inhibited by 42 per cent, 69 per cent and at the maturity inhibited by 43 per cent, 48 per cent respectively as compared to control.

When UV-B exposed plants were sprayed along with IAA (10^{-7} M) concentration, the mitigatory response was observed in terms of the fresh weight and dry weight at the 75th day stage of growth up to maturity and increased by ca. 44 per cent and 82 per cent respectively as compared to UV-B only. When UV-B exposure was given along with Kn (10^{-5}) concentration, the promotory effect was showed at the 90th day stage of growth and increased by ca. 24 per cent and 13 per cent respectively as compared to UV-B (3 hrs. daily) radiation only. When UV-B exposed plants were also sprayed along with GA_3 (10^{-6} M) concentration, the similar result was found at the maturity stage of growth and increased by ca. 19 per cent and 20 per cent respectively as compared to individual treatment of UV-B radiation (3 hrs. daily). Table 4.5.

Brassica juncea PR-15

Growth pattern of *Brassica juncea PR-15* (Rai) was also studied under influence of UV-B radiation, individually and in combination of certain plant growth regulator (PGRs) in terms of the stem, leaf, root, flower and fruit development. During field study in control, the value of stem length, fresh weight and dry weight was found maximum at the 15th day stage of growth and recorded as ca. 4.85 cm/pl, 0.079 and 0.035 g/pl respectively and continuously increased up to maturity and amounted as 40.25 cm/pl, 0.79 and 0.55 g/pl respectively. When the plot was subjected to UV-B only (3 hrs. daily), a reduction was found in all these parameters as compared to the control plot. The maximum inhibition in the length was noticed at the 45th day stage of growth and reduced by ca. 19 per cent as compared to control plot. The maximum deleterious effects on fresh and dry weight of stem were observed at the 90th day stage of growth and inhibited by ca. 30 per cent and 40 per cent respectively as compared to the control. When UV-B exposure was given in combination of some plant growth regulators (PGRs), a promotory effects was recorded in all these parameters as compared to individual treatment of UV-B radiation.

The IAA (10^{-7} M) concentration was found to be most promotory effective, against UV-B induced deleterious impacts in all the considered parameters. The maximum promotory effects were showed in the length and observed at the 45th day stage of growth and promoted by ca. 19 per cent over the UV-B treatment. The maximum fresh weight of stem was found at the maturity stage of growth and increased by ca. 18 per cent as compared to UV-B exposed plot. When the UV-B radiation was given along with Kn (10^{-5} M) concentration, a maximum promotory effects was noticed in terms of

Table 4.5: Fruit growth patterns of *Brassica compestris PT-303* as affected by UV-B radiation (3 hrs. daily), individually and in combination of IAA, Kn and GA_3.

Treat-ments	*Para-meters*	*Crop Age in Days*								
		15	*30*	*45*	*60*	*75*	*90*	*105*	*120*	*135*
A	F.W.(g)	–	–	–	–	**0.058**±0.0049	**0.163**±0.0050	**0.289**±0.079	**0.384**±0.0049	**0.578**±0.004
	D.W.(g)	–	–	–	–	**0.034**±0.0075	**0.043**±0.0055	**0.144**±0.001	**0.242**±0.0038	**0.342**±0.004
B	F.W.(g)	–	–	–	–	**0.029**±0.0014	**0.078**±0.005	**0.169**±0.003	**0.242**±0.005	**0.387**±0.003
	D.W.(g)	–	–	–	–	**0.014**±0.0003	**0.035**±0.037	**0.044**±0.004	**0.121**±0.007	**0.245**±0.004
C	F.W.(g)	–	–	–	–	**0.042**±0.0039	**0.096**±0.004	**0.179**±0.007	**0.284**±0.0041	**0.487**±0.076
	D.W.(g)	–	–	–	–	**0.021**±0.0019	**0.040**±0.053	**0.068**±0.003	**0.142**±0.003	**0.293**±0.003
D	F.W.(g)	–	–	–	–	**0.043**±0.0029	**0.097**±0.003	**0.186**±0.008	**0.289**±0.005	**0.487**±0.005
	D.W.(g)					**0.029**±0.0030	**0.038**±0.006	**0.068**±0.071	**0.144**±0.002	**0.283**±0.001
E	F.W.(g)	–	–	–	–	**0.048**±0.0059	**0.098**±0.002	**0.187**±0.006	**0.289**±0.025	**0.489**±0.026
	D.W.(g)	–	–	–	–	**0.029**±0.0091	**0.039**±0.004	**0.073**±0.043	**0.146**±0.035	**0.254**±0.065

Table 4.6: Stem growth patterns of field grown *Brassica juncea PR-15* as affected by UV-B radiation (3 hrs. daily), individually and in combination of IAA, Kn and GA_3.

Treat-ments	*Para-meters*	*Crop Age in Days*								
		15	*30*	*45*	*60*	*75*	*90*	*105*	*120*	*135*
A	Length (cm)	4.85±0474	7.45±0.955	12.20±0.0.537	16.50±0.685	19.00±1.005	22.50±0.971	24.20±0.737	29.20±0.977	40.25±0.965
	F.W. (g)	0.079±0.0069	0.123±0.029	0.269±0.0061	0.417±0.053	0.519±0.089	0.723±0.0031	0.725±0.0071	0.727±0.069	0.788±0.0054
	D.W. (g)	0.034±0.0022	0.100±0.025	0.135±0.095	0.254±0.044	0.315±0.0071	0.412±0.039	0.512±0.0079	0.514±0.0072	0.544±0.0059
B	Length (cm)	3.95±0.598	5.95±0.685	9.90±1.619	13.80±0.421	15.45±0.437	19.80±3.735	22.18±3.125	25.60±3.186	35.95±3.115
	F.W.(g)	0.055±0.0012	0.098±0.0024	0.248±0.005	0.378±0.087	0.383±0.039	0.512±0.025	0.514±0.097	0.564±0.069	0.645±0.099
	D.W.(g)	0.024±0.0015	0.056±0.0011	0.114±0.069	0.214±0.080	0.243±0.0073	0.311±0.0049	0.312±0.096	0.332±0.035	0.403±0.0034
C	Length (cm)	4.15±0.411	6.90±1.264	11.80±0.632	14.05±0.895	17.85±1.453	20.80±1.229	23.90±1.100	26.70±1.059	38.96±4.62
	F.W. (g)	0.066±0.0024	0.108±0.098	0.218±0.079	0.399±0.0081	0.429±0.069	0.622±0.0091	0.683±0.085	0.687±0.078	0.768±0.0058
	D.W.(g)	0.031±0.021	0.075±0.0015	0.115±0.033	0.224±0.0075	0.286±0.0025	0.351±0.0072	0.341±0.079	0.355±0.069	0.505±0.0068
D	Length (cm)	4.30±0.349	6.75±0.950	11.86± 0.564	14.05±0.497	17.45±1.832	20.70±4.467	23.00±1.030	26.60±2.611	38.85±4.690
	F.W. (g)	0.066±0.0023	0.105±0.035	0.218±0.045	0.225±0.035	0.397±0.0072	0.622±0.0061	0.682±0.0092	0.686±0.059	0.767±0.095
	D.W. (g)	0.032±0.0024	0.074±0.023	0.104±0.075	0.222±0.0071	0.274±0.082	0.351±0.096	0.341±0.0039	0.343±0.048	0.525±0.0039
E	Length (cm)	4.40±0.210	6.85±0.483	11.90±0.658	14.50±1.065	17.75±0.790	20.60±3.871	23.20±0.816	26.70±1.068	38.80±5.995
	F.W. (g)	0.066±0.055	0.109±0.057	0.228±0.079	0.398±0.045	0.473±0.055	0.652±0.029	0.655±0.055	0.659±0.051	0.769±0.087
	D.W. (g)	0.032±0.050	0.075±0.047	0.124±0.042	0.224±0.049	0.291±0.070	0.361±0.052	0.334±0.050	0.357±0.084	0.505±0.0057

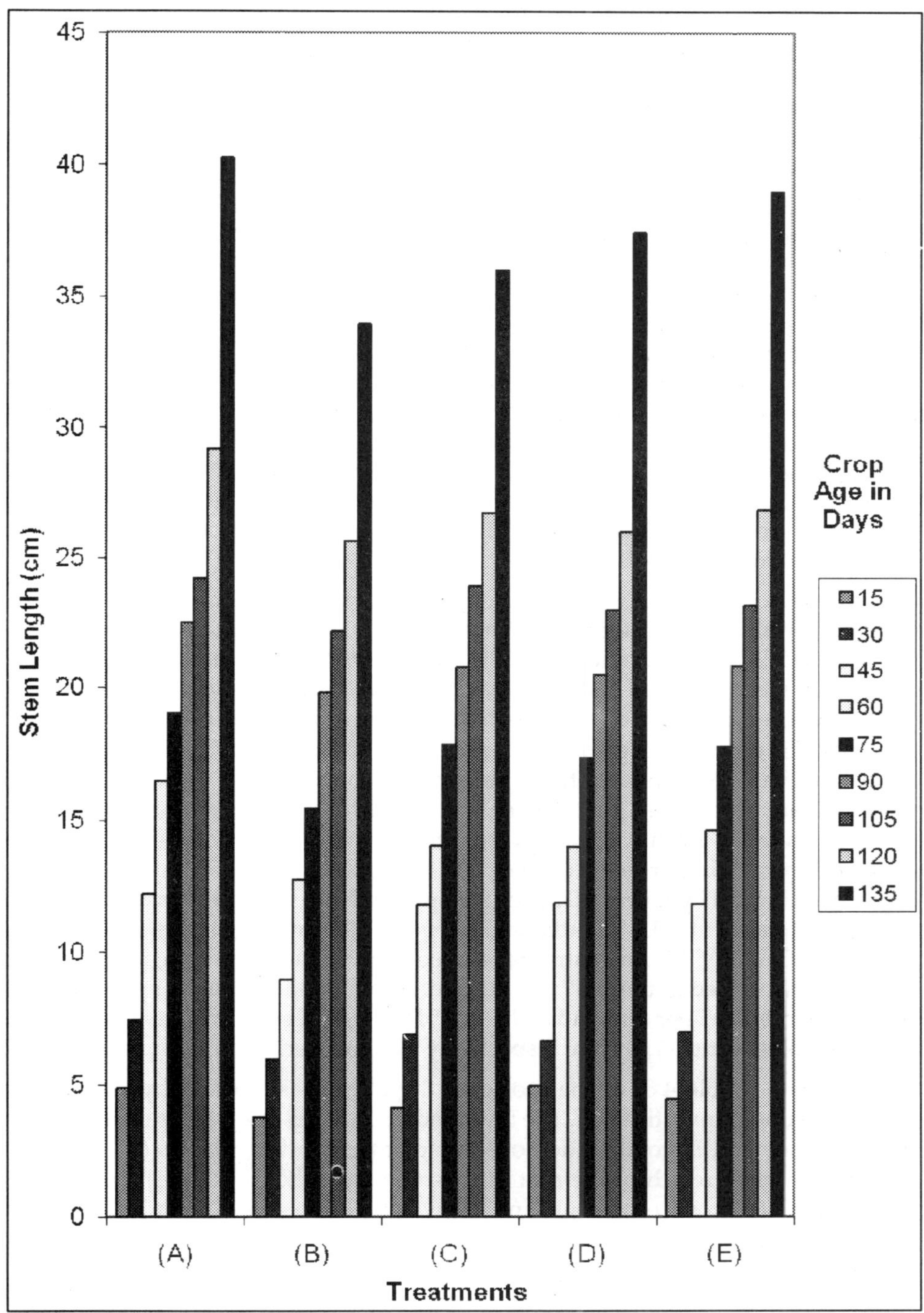

Figure 4.4: Stem growth patterns of field grown *Brassica juncea PR-15* as affected by different treatments.

length, fresh and dry weight at the 45th day stage of growth and increased by ca. 19 per cent, 21 per cent and 15 per cent as compared to UV-B exposure only. When UV-B treatment was also given along with GA_3 (10^{-7} M) concentration, the maximum promotion in length, fresh weight and dry weight were observed at the 45th day stage of growth and increased by ca. 20 per cent, 11 per cent and 24 per cent respectively as compared to the individual treatment of UV-B exposure (Table 4.6 and Figure 4.4).

Table 4.7 and Figure 4.5, had shown the data of the leaf growth pattern as affected by various treatments with and without UV-B, along with some plant growth regulators (PGRs). The value of leaf area growth pattern (cm^2/pl), fresh weight and dry weight (g/pl) of leaf was recorded maximum at the 15th day stage of growth and amounted as ca. 3.97 (cm^2/pl), 0.138 (g/pl) and 0.118 (g/pl) respectively and found to be increased continuously up to maturity and recorded 34.15 (cm^2/pl), 0.375 (g/pl) and 0.243 (g/pl) respectively. When the plants were exposed to UV-B radiation only, a reduction was observed in the leaf area, fresh weight, and dry weight of leaf as compared to the control. The maximum reduction in the leaf area was observed at the 45th day stage of growth and recorded as ca. 72 per cent. A significant reduction of fresh weight and dry weight of leaf were noticed at the 60th day and 75th day stage of growth and reduced by ca. 10 per cent, 21 per cent 5 per cent and 9 per cent respectively as compared to the control.

When UV-B exposure was given along with some PGRs, the mitigatory response was found in all these considered parameters. When UV-B exposure was given along with IAA (10^{-7}M) concentration, a maximum value of the leaf area, fresh weight and dry weight were recorded at the 45th day and increased by ca. 16 per cent, 16.5 per cent, 16 per cent; at the 60th day increased by ca. 13.8 per cent, 4.8 per cent, 7.8 per cent; and at the 75th day stage of growth increased by ca. 13.9 per cent, 5.2 per cent, 10.3 per cent respectively, as compared to the individual treatment of UV-B exposure. When the UV-B radiation was supplied along with Kn (10^{-5}M) concentration, a maximum value of leaf area, fresh weight and dry weight were noticed at the 45th day and increased by ca. 16.6 per cent, 16.5 per cent, 34 per cent : at the 75th day increased by ca.12.7 per cent, 5.17 per cent, 10 per cent respectively up to maturity as compared to individual treatment of UV-B radiation. When the UV-B radiation was also supplied along with GA_3 (10^{-7} M) concentration, a maximum value of leaf area, fresh and dry weight were noticed at the 105th day stage of growth and increased by ca. 10.3 per cent, 3.6 per cent, 35 per cent and at the 120th day stage of growth increased by ca. 10 per cent, 3.3 per cent, 34 per cent respectively as compared to UV-B exposure only.

The data (Table 4.8) of the root growth patterns as affected by the various treatments under investigation. The UV-B radiation (3 hrs. daily) was given daily to the plot alone and in combination of some plant growth regulators (PGRs) effect significantly the growth of the root of *Brassica juncea PR-15* (Rai). The maximum deleterious effects on fresh weight and dry weight of roots were recorded at the 30th and 60th day stage of growth and decreased by ca. 20 per cent, 62 per cent and ca. 52 per cent, 62 per cent respectively as compared to the control. When UV-B irradiation was given along with PGRs, a promotory effects was observed in terms of concerned parameters.

Table 4.7: Leaf growth patterns of field grown *Brassica juncea PR-15* as affected by UV-B radiation (3 hrs. daily), individually and in combination of IAA, Kn and GA_3.

Treat-ments	Para-meters	Crop Age in Days								
		15	30	45	60	75	90	105	120	135
A	Leaf Area (cm^2)	3.96±0.64	6.82±0.35	12.05±0.65	14.85±0.81	18.42±0.70	21.80±0.57	23.21±2.81	25.15±1.02	34.15±4.30
	F.W. (g)	0.138±0.029	0.124±0.0069	0.129±0.073	0.213±0.059	0.219±0.0063	0.296±0.0063	0.313±0.075	0.318±0.073	0.375±0.047
	D.W. (g)	0.118±0.0039	0.053±0.0038	0.076±0.047	0.141±0.048	0.116±0.0043	0.163±0.0038	0.211±0.039	0.213±0.064	0.243±0.038
B	Leaf Area (cm^2)	2.85±0.707	5.21±0.869	9.64±1.13	11.64±0.73	14.85±0.68	17.83±0.63	19.75±4.44	22.65±3.14	30.43± 5.97
	F.W. (g)	0.094±0.0069	0.099±0.0090	0.105±0.008	0.198±0.0069	0.209±0.0069	0.264±0.0071	0.278±0.043	0.283±0.037	0.293±0.067
	D.W. (g)	0.042±0.0073	0.039±0.0061	0.053±0.009	0.112±0.0038	0.095±0.0074	0.132±0.0072	0.144±0.053	0.149±0.023	0.182±0.059
C	Leaf Area (cm^2)	2.92±0.706	6.02±0.86	11.25±1.96	13.25±1.13	16.92±0.87	19.80±4.39	21.80±1.00	24.00±0.54	32.10±4.50
	F.W. (g)	0.114±0.0091	0.119±0.065	0.123±0.063	0.203±0.076	0.209±0.0054	0.273±0.0048	0.288±0.050	0.293±0.0054	0.298±0.0070
	D.W. (g)	0.082±0.029	0.045±0.039	0.061±0.097	0.121±0.037	0.105±0.0063	0.141±0.037	0.189±0.049	0.192±0.0034	0.214±0.069
D	Leaf Area (cm^2)	3.00±0.86	6.01±0.85	11.24±2.52	13.15±1.17	16.74±0.81	19.57±4.27	21.75±0.98	23.81±0.49	32.14±4.50
	F.W. (g)	0.114±0.0092	0.116±0.0065	0.122±0.038	0.203±0.0083	0.209±0.0064	0.263±0.063	0.287±0.061	0.289±0.084	0.298±0.054
	D.W. (g)	0.062±0.0071	0.048±0.059	0.071±0.035	0.121±0.047	0.105±0.0039	0.142±0.045	0.187±0.051	0.185±0.074	0.213±0.035
E	Leaf Area (cm^2)	3.04±0.0665	6.00±0.77	11.25±2.65	13.31±1.14	16.44±0.92	19.71±4.33	21.79±0.95	24.92±3.10	32.25±4.57
	F.W. (g)	0.114±0.025	0.119±0.081	0.123±0.061	0.203±0.069	0.209±0.039	0.273±0.075	0.288±0.050	0.292±0.079	0.298±0.036
	D.W. (g)	0.084±0.031	0.046±0.073	0.072±0.0051	0.121±0.0073	0.129±0.0048	0.142±0.063	0.194±0.047	0.191±0.038	0.216±0.031

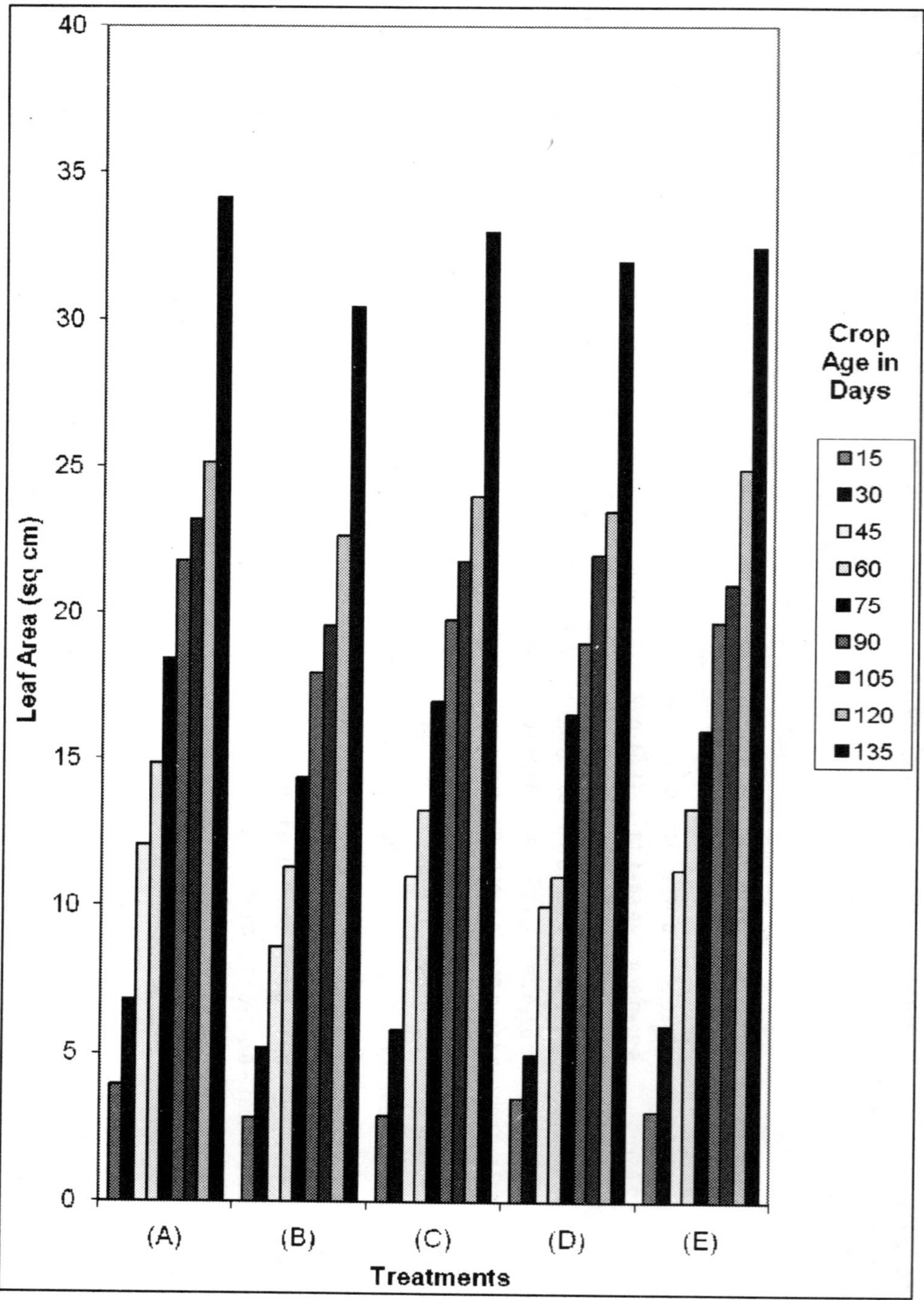

Figure 4.5: Leaf growth patterns of field grown *Brassica juncea PR-15* as affected by different treatments.

Table 4.8: Root growth patterns of field grown *Brassica juncea PR-15* as affected by UV-B radiation (3 hrs. daily), individually and in combination of IAA, Kn and GA_3.

Treat-ments	*Para-meters*	*Crop Age in Days*								
		15	*30*	*45*	*60*	*75*	*90*	*105*	*120*	*135*
A	F.W.(g)	**0.079**±0.007	**0.121**±0.073	**0.124**±0.069	**0.262**±0.001	**0.415**±0.031	**0.419**±0.008	**0.518**±0.009	**0.713**±0.073	**1.325**±0.005
	D.W.(g)	**0.033**±0.003	**0.111**±0.006	**0.114**±0.007	**0.234**±0.039	**0.310**±0.069	**0.314**±0.063	**0.414**±0.004	**0.511**±0.006	**0.922**±0.006
B	F.W.(g)	**0.067**±0.002	**0.098**±0.009	**1.122**±0.003	**0.129**±0.029	**0.277**±0.063	**0.351**±0.009	**0.397**±0.008	**0.513**±0.006	**0.944**±0.009
	D.W.(g)	**0.021**±0.096	**0.043**±0.006	**0.095**±0.005	**0.113**±0.069	**0.234**±0.083	**0.271**±0.003	**0.214**±0.061	**0.313**±0.007	**0.723**±0.002
C	F.W.(g)	**0.072**±0.078	**0.112**±0.035	**0.119**±0.009	**0.257**±0.073	**0.352**±0.069	**0.398**±0.007	**0.497**±0.054	**0.602**±0.005	**1.246**±0.069
	D.W.(g)	**0.030**±0.068	**0.053**±0.065	**0.104**±0.076	**0.213**±0.068	**0.290**±0.082	**0.291**±0.005	**0.351**±0.064	**0.451**±0.009	**0.823**±0.003
D	F.W.(g)	**0.071**±0.083	**0.111**±0.007	**0.118**±0.035	**0.255**±0.034	**0.362**±0.008	**0.398**±0.004	**0.486**±0.063	**0.602**±0.007	**1.268**±0.073
	D.W.(g)	**0.031**±0.079	**0.063**±0.068	**0.103**±0.043	**0.213**±0.006	**0.281**±0.007	**0.292**±0.003	**0.352**±0.053	**0.451**±0.003	**0.822**±0.008
E	F.W.(g)	**0.072**±0.005	**0.112**±0.079	**0.117**±0.032	**0.256**±0.083	**0.371**±0.064	**0.397**±0.053	**0.487**±0.063	**0.622**±0.005	**1.298**±0.059
	D.W.(g)	**0.032**±0.006	**0.053**±0.009	**0.103**±0.041	**0.213**±0.035	**0.291**±0.076	**0.292**±0.043	**0.341**±0.078	**0.461**±0.003	**0.824**±0.008

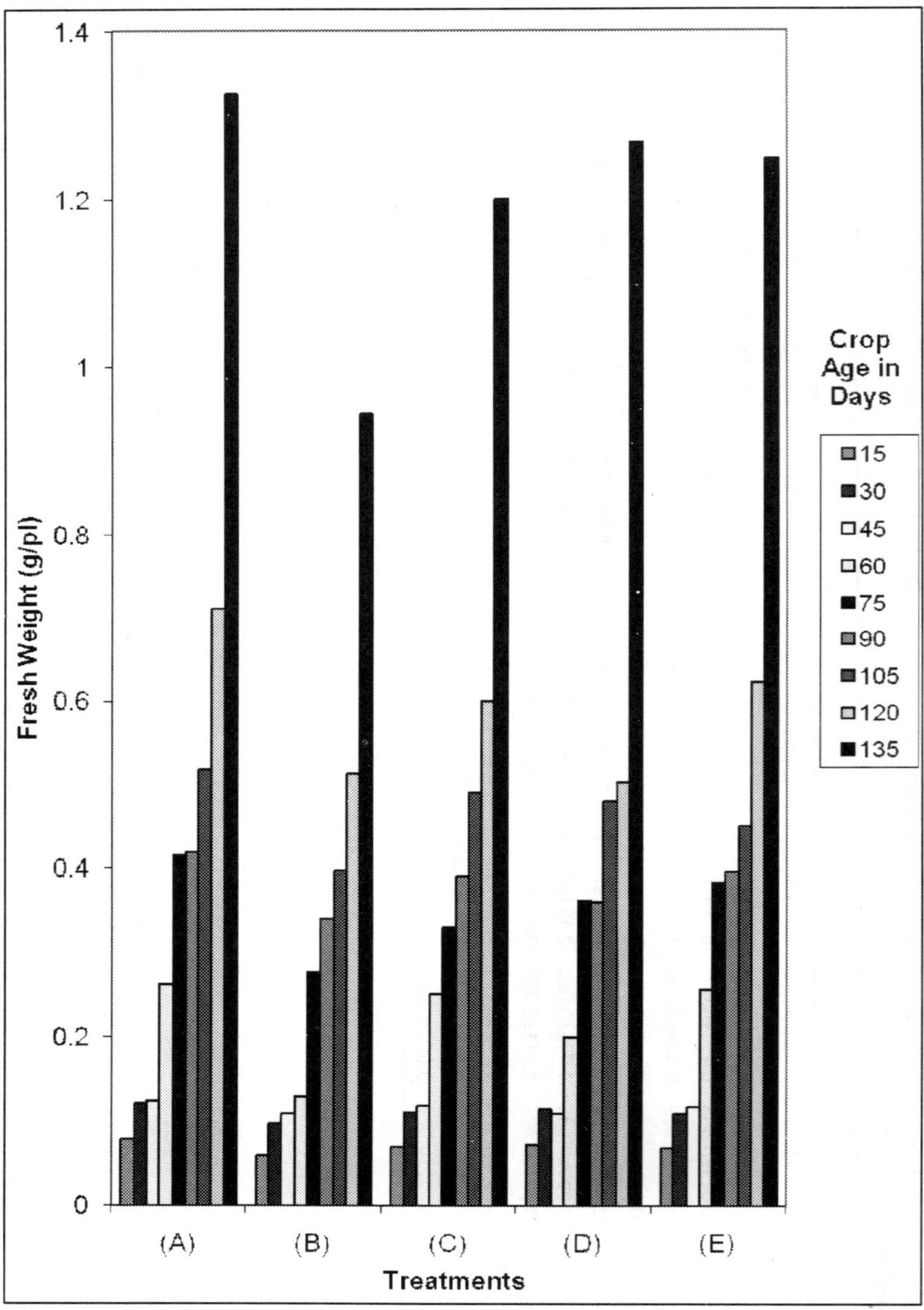

Figure 4.6: Root growth patterns of field grown *Brassica juncea PR-15* as affected by different treatments.

When UV-B exposure was given along with IAA (10^{-7}M) concentration, the promotory effects were reported in terms of fresh weight and dry weight at the 15th day and 75th day stage of growth and increased by ca. 6.5 per cent, 42 per cent and ca. 26.9 per cent, 23.9 per cent respectively as compared to UV-B exposed plot only. When UV-B treatment was given along with Kn (10^{-5} M) concentration, a mitigatory effects was recorded in terms of the fresh weight and dry weight of the root at the 15th day and 75th day stage of growth and increased by ca. 5.6 per cent, 51 per cent and ca. 30.6 per cent, 20 per cent respectively as compared to individual treatment of UV-B radiation. When UV-B radiation was also given along with GA_3 (10^{-7} M) concentration, a mitigatory response was also noticed in terms of the fresh and dry weight of root recorded at the 30th day and increased by ca. 14 per cent, 24 per cent; at the 75th day increased by ca. 34 per cent, 24.2 per cent and at the 120th day stage of growth up to maturity and increased by ca. 21.2 per cent, 47 per cent respectively as compared to individual treatment of UV-B radiation (3 hrs. daily).

UV-B radiation was reported to induce deleterious impacts on the growth of flowers in terms of fresh and dry weight. In the present study, the flower analysis was carried out from the 60th day stage of growth up to maturity in terms of fresh and dry weight (g/plant) and data were presented in the (Table 4.9). When UV-B exposure (3 hrs. daily) was supplied individually, exhibited inhibition in all these considered parameters as compared to control. The maximum deleterious impacts on fresh and dry weight of flowers were recorded from the 60th day and inhibited by ca. 45 per cent, 68 per cent; at the 105th day inhibited by ca. 25.6 per cent, 21 per cent and maturity stage of growth inhibited by ca. 20.3 per cent, 24 per cent respectively as compared to control plot. When UV-B exposed plants were treated along with IAA (10^{-7} M), Kn (10^{-5} M) and GA_3 (10^{-7} M) concentrations, the promotion was observed in all these considered parameters.

When UV-B exposed plants were sprayed along with IAA (10^{-7} M) concentration, a mitigatory response was recorded in terms of the fresh weight and dry weight of the flowers at the 60th day stage of growth up to maturity and increased by ca. 55 per cent, 90 per cent and ca. 17 per cent, 5.8 per cent respectively as compared to individual treatments of UV-B radiation. When UV-B exposed plants were sprayed with Kn (10^{-5} M) concentration, a mitigatory response was noted in terms of the fresh weight and dry weight at the 60th day stage of growth up to maturity and increased by ca. 30 per cent, 90 per cent and 16.6 per cent, 11.7 per cent respectively as compared to UV-B treatment only. When UV-B exposed plants were also supplied with GA_3 (10^{-7} M) concentration, a promotory effects was identified in terms of the fresh and dry weight of flowers as compared to individual treatment of UV-B radiation (3 hrs. daily).

Fruiting determine the yield of crop in the agriculture system, therefore, the excellence of fruiting was also studied. An increase in fresh weight and dry weight of fruits was observed when sprayed in combination of different plant growth regulators (PGRs), as compared to over UV-B (3 hrs. daily) radiation. Maximum decrease in fresh weight and dry weight have been observed at the 75th day and 90th day stage of the growth and inhibited by ca. 46 per cent, 48.4 per cent and ca. 42 per cent, 85.7 per cent respectively as compared to control plot.

Table 4.9: Flower growth patterns of *Brassica juncea PR-15* as affected by UV-B radiation (3 hrs. daily), individually and in combination of IAA, Kn and GA_3.

Treat-ments	*Para-meters*	*Crop Age in Days*								
		15	*30*	*45*	*60*	*75*	*90*	*105*	*120*	*135*
A	F.W. (g)	–	–	–	**0.058**±0.005	**0.071**±0.0004	**0.079**±0.005	**0.087**±0.004	**0.093**±0.005	**0.112**±0.006
	D.W. (g)	–	–	–	**0.035**±0.004	**0.048**±0.0049	**0.053**±0.0051	**0.063**±0.0092	**0.065**±0.0054	**0.083**±0.006
B	F.W. (g)	–	–	–	**0.032**±0.004	**0.052**±0.054	**0.061**±0.009	**0.065**±0.005	**0.076**±0.007	**0.083**±0.0035
	D.W. (g)	–	–	–	**0.011**±0.025	**0.039**±0.0039	**0.039**±0.0054	**0.049**±0.0052	**0.052**±0.0037	**0.063**±0.0020
C	F.W. (g)	–	–	–	**0.048**±0.005	**0.062**±0.003	**0.066**±0.002	**0.073**±0.039	**0.079**±0.021	**0.097**±0.023
	D.W. (g)	–	–	–	**0.021**±0.0009	**0.042**±0.007	**0.044**±0.009	**0.052**±0.023	**0.053**±0.0031	**0.071**±0.0035
D	F.W. (g)	–	–	–	**0.041**±0.005	**0.053**±0.003	**0.062**±0.072	**0.068**±0.0041	**0.078**±0.0041	**0.097**±0.054
	D.W. (g)				**0.022**±0.0031	**0.042**±0.003	**0.043**±0.069	**0.053**±0.0025	**0.053**±0.0023	**0.072**±0.036
E	F.W. (g)	–	–	–	**0.041**±0.0052	**0.053**±0.004	**0.062**±0.053	**0.068**±0.079	**0.078**±0.0059	**0.097**±0.0044
	D.W. (g)	–	–	–	**0.023**±0.0097	**0.043**±0.007	**0.044**±0.025	**0.043**±0.069	**0.053**±0.069	**0.065**±0.037

Table 4.10: Fruit growth patterns of *Brassica juncea-PR-15* as affected by UV-B radiation (3 hrs. daily), individually and in combination of IAA, Kn and GA_3.

Treat-ments	Para-meters	Crop Age in Days								
		15	30	45	60	75	90	105	120	135
A	F.W.(g)	–	–	–	–	0.064±0.0054	0.164±0.0050	0.289±0.007	0.385±0.0048	0.579±0.003
	D.W.(g)	–	–	–	–	0.032±0.0060	0.042±0.0051	0.143±0.002	0.242±0.0038	0.342±0.004
B	F.W.(g)	–	–	–	–	0.029±0.0014	0.069±0.004	0.169±0.003	0.124±0.006	0.387±0.002
	D.W.(g)	–	–	–	–	0.015±0.002	0.035±0.0038	0.045±0.005	0.141±0.008	0.245±0.004
C	F.W.(g)	–	–	–	–	0.041±0.0040	0.097±0.004	0.179±0.007	0.283±0.0042	0.498±0.0075
	D.W.(g)	–	–	–	–	0.020±0.0019	0.039±0.005	0.076±0.003	0.192±0.0039	0.299±0.0031
D	F.W.(g)	–	–	–	–	0.042±0.0030	0.098±0.0034	0.186±0.008	0.289±0.026	0.488±0.006
	D.W.(g)					0.021±0.049	0.039±0.0009	0.076±0.061	0.186±0.013	0.299±0.002
E	F.W.(g)	–	–	–	–	0.048±0.0036	0.097±0.0021	0.187±0.006	0.289±0.0026	0.448±0.027
	D.W.(g)	–	–	–	–	0.024±0.0049	0.039±0.0003	0.078±0.048	0.194±0.0036	0.298±0.055

When UV-B exposed plants were sprayed along with IAA (10^{-7} M) concentration, it showed maximum value of fresh weight and dry weight of fruit at the 75th day and 90th day stage of growth and increased by ca. 38.9 per cent, 34.1 per cent and ca. 40.5 per cent, 11.4 per cent respectively as compared to individual treatment of UV-B radiation (3 hrs. daily). When UV-B exposure was given along with Kn (10^{-5} M) concentration, the maximum increase in the fresh weight and dry weight of fruit was reported at the 90th and 105th day stage of growth and increased by ca. 42.2 per cent, 12.8 per cent and ca. 10.3 per cent, 68 per cent respectively as compared to UV-B exposure. When UV-B irradiated plants were also supplied along with GA_3 (10^{-7} M) concentration, the maximum increase in fresh weight and dry weight of fruit was noted at the 105th day stage of growth and increased by ca. 10.6 per cent, 73 per cent; at the 120th day increased by ca. 18.9 per cent, 37.1 per cent and up to maturity increased by ca. 26 per cent, 21 per cent respectively as compared to individual treatment of UV-B radiation (3 hrs. daily). Table 4.10.

Chapter 5

UV-B Irradiance Induced Deleterious Effects Individually and in Combination of Different Plant Growth Regulators (PGRs) on the Productivity and Biomass Partitioning

Deleterious effects of UV-B irradiance alone and along with IAA (10^{-7} M), Kn (10^{-5} M) and GA_3 (10^{-6} M) in *Brassica compestris PT-303* and (10^{-7}M) in *Brassica juncea PR-15* were studied on the standing crop, net primary productivity and yield attributes studied in the *Brassica compestris PT-303* and *Brassica juncea PR-15* respectively. The samples were collected for the studies on the basis of morphological resemblance and standing crop was calculated on the per plant basis as serve to describe in the earlier studies.

Standing Crop

Data depicted in Tables 5.1 and 5.2 and Figures 5.1 and 5.2, exhibited the amount of standing crops of different parts of the plant *viz.* root, stem, leaf, flower and fruits were observed to increase linearly with the forward movement in the growth stage but at the maturity stage the leaves were observed to reduce due to the leaf fall in both varieties of the mustard crops. In the Table 5.1 showed the response of UV-B radiation

Table 5.1: Standing crop (g/plant), on the dry weight basis of field grown *Brassica compestris PT-303* as affected by UV-B radiation, individually and in combination of IAA, Kn and GA_3.

Treat-ments	*Para-meters*	*Crop Age in Days*								
		15	*30*	*45*	*60*	*75*	*90*	*105*	*120*	*135*
A	Root	0.035	0.091	0.109	0.234	0.310	0.314	0.413	0.511	0.822
	Stem	0.034	0.102	0.134	0.315	0.414	0.513	0.512	0.514	0.544
	Leaf	0.084	0.091	0.106	0.112	0.156	0.212	0.214	0.311	0.315
	Flower	–	–	–	0.034	0.048	0.053	0.058	0.062	0.073
	Fruit	–	–	–	–	0.034	0.043	0.145	0.242	0.342
	Total	**0.153**	**0.284**	**0.349**	**0.695**	**0.962**	**1.135**	**1.342**	**1.641**	**2.098**
B	Root	0.030	0.043	0.095	0.098	0.232	0.271	0.214	0.312	0.622
	Stem	0.024	0.055	0.104	0.254	0.244	0.313	0.320	0.332	0.323
	Leaf	0.052	0.054	0.074	0.142	0.114	0.142	0.142	0.213	0.235
	Flower	–	–	–	0.013	0.036	0.039	0.044	0.053	0.066
	Fruit	–	–	–	–	0.019	0.035	0.045	0.121	0.245
	Total	**0.106**	**0.152**	**0.273**	**0.507**	**0.645**	**0.800**	**0.765**	**1.031**	**1.491**
C	Root	0.032	0.053	0.103	0.223	0.280	0.301	0.321	0.412	0.724
	Stem	0.032	0.057	0.124	0.264	0.276	0.341	0.341	0.354	0.494
	Leaf	0.064	0.069	0.081	0.152	0.125	0.151	0.164	0.251	0.259
	Flower	–	–	–	0.022	0.029	0.039	0.052	0.059	0.069
	Fruit	–	–	–	–	0.021	0.040	0.068	0.142	0.293
	Total	**0.128**	**0.179**	**0.308**	**0.661**	**0.731**	**0.872**	**0.946**	**1.218**	**1.836**
D	Root	0.030	0.052	0.102	0.213	0.291	0.302	0.322	0.411	0.723
	Stem	0.030	0.055	0.124	0.265	0.274	0.312	0.341	0.352	0.524
	Leaf	0.064	0.058	0.082	0.152	0.125	0.151	0.174	0.253	0.257
	Flower	–	–	–	0.022	0.044	0.047	0.043	0.054	0.065
	Fruit	–	–	–	–	0.029	0.038	0.068	0.144	0.283
	Total	**0.124**	**0.165**	**0.308**	**0.652**	**0.763**	**0.850**	**0.948**	**1.216**	**1.852**
E	Root	0.032	0.053	0.103	0.223	0.291	0.298	0.331	0.420	0.724
	Stem	0.025	0.056	0.124	0.264	0.281	0.321	0.351	0.362	0.525
	Leaf	0.064	0.059	0.081	0.153	0.126	0.151	0.174	0.262	0.286
	Flower	–	–	–	0.022	0.044	0.048	0.053	0.055	0.064
	Fruit	–	–	–	–	0.029	0.039	0.073	0.146	0.254
	Total	**0.121**	**0.168**	**0.308**	**0.662**	**0.771**	**0.857**	**0.982**	**1.245**	**1.853**

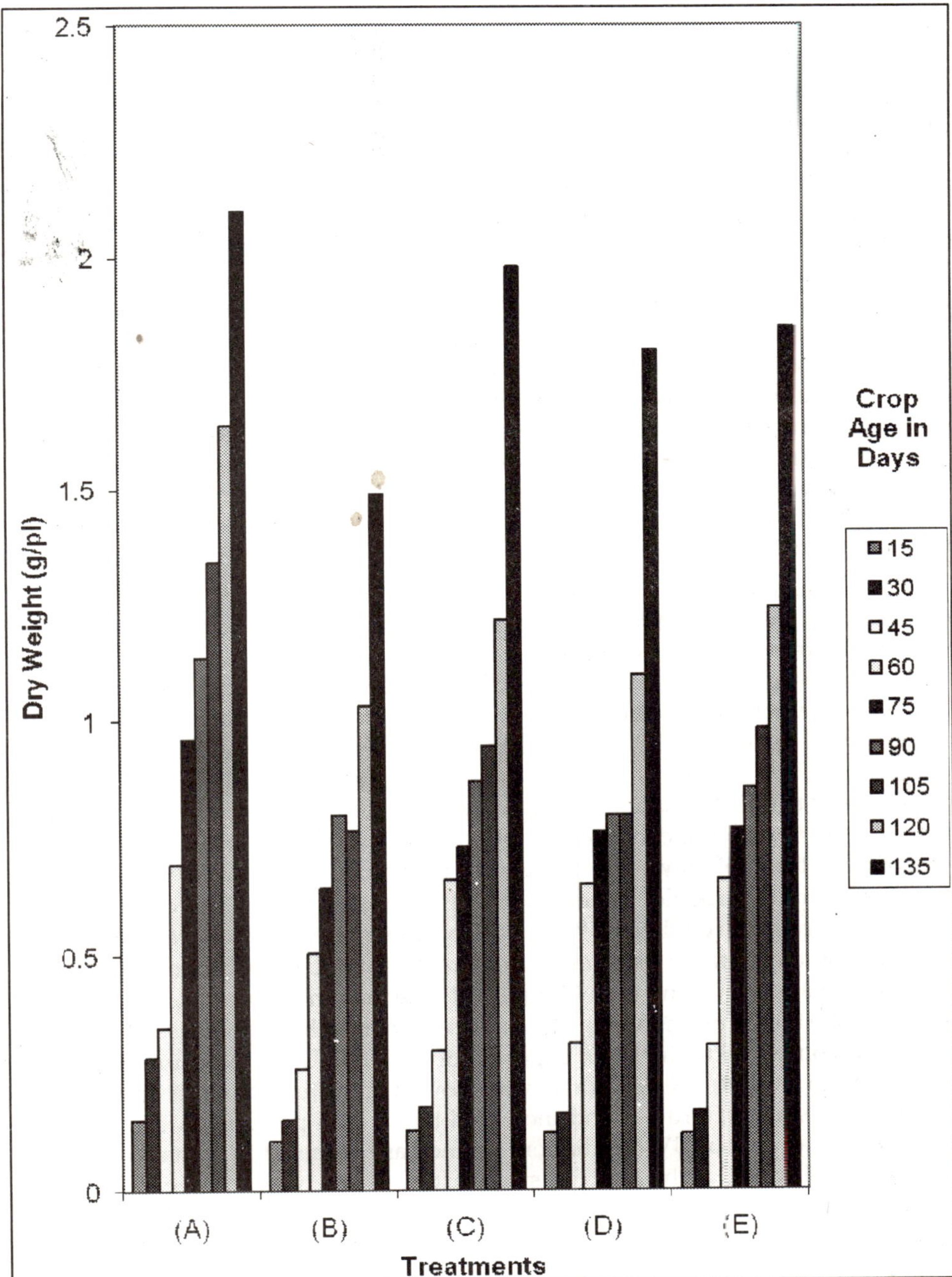

Figure 5.1: Total standing crop (g/plant), on the dry weight basis of field *Brassica compestris PT-303* as affected by different treatments.

individually and in combination of IAA (10^{-7}M), Kn (10^{-5}M) and GA_3 (10^{-6}M) on the standing crop of *Brassica compestris PT-303*.

During field study, in control plot the total biomass (Root + stem + Leaf + flower + fruits) was calculated and found maximum at the 15th day stage of growth and recorded as 0.153 g/pl. These values were observed to increase up to maturity and amounted as 0.283, 0.345, 0.694, 0.693, 1.621, 1.336, 1.639 and 2.098 g/pl at the 30th, 45th, 60th, 75th, 90th, 105th, 120th and 135th day stage respectively. The plot B, which was subjected to only UV-B exposure (3 hrs. daily), the maximum deleterious impacts was observed in the total biomass (Root + stem + Leaf + Flower + Fruit), at the 105th day stage of growth and deceased by ca. 45 per cent as compared to control. The value of biomass was found to be increased, when UV-B exposure was given along with IAA (10^{-7} M) and increased by ca. 26.8 per cent at the 105th day stage of the growth as compared to individual treatment of UV-B radiation (3 hrs. daily). When UV-B exposed plants were sprayed along with Kn (10^{-5}M) concentration, a maximum value of total biomass was recorded at the 105th day stage of growth and increased by ca. 27.3 per cent as compared to the UV-B exposure only. When UV-B exposed plants were also supplied along with GA_3 (10^{-6} M), the maximum value of total biomass was recorded at the 105th day stage of the growth and increased by ca. 30.4 per cent as compared to the UV-B (3 hrs. daily) exposure only.

The perusal of data shown in Table 5.2 present standing crop of various plant parts *viz.* root, stem, leaf, flower and fruits of *Brassica juncea PR-15* as affected by the different treatments. In the control plot, the total biomass (root + stem + leaf + flower + fruits) was recorded minimum at the 15th day stage of growth and noticed as 0.178 g/pl. These values were found to increase continuously and linearly up to maturity and observed to be ca. 0.263, 0.665, 0.822, 1.088, 1.243, 1.546 and 2.135 g/pl at the 30th, 45th, 60th,75th, 90th, 105th and 135th day respectively. When the plants of plot B were exposed to UV-B radiation only, it was observed deleterious to the total biomass at all the growth stages as compared to the control. The maximum deleterious effects was found at the 105th day stage of the growth and decreased by ca. 39 per cent as compared to the control plot. When UV-B radiation was given in the combination of certain plant growth regulators (PGRs), the promotory effects were reported in the total biomass as compared to the individual treatment of UV-B radiation. When UV-B radiation was given along with IAA (10^{-7}M), the maximum promotory effect was showed in the total biomass at the 105th day stage of growth and increased by ca. 32.4 per cent as compared to the UV-B exposure only. When UV-B exposure was given along with Kn (10^{-5} M), the promotion was showed in the total biomass and noted at the 105th day stage of growth and increased by ca. 31.4 per cent as compared to individual treatment of UV-B radiation. An increase of the total biomass of ca. 30.8 per cent was observed at the 105th day stage of the growth, when UV-B exposure was given along with GA_3 (10^{-7} M) concentration as compared to individual treatment of UV-B radiation (3 hrs. daily).

Net Production (g/pl)

Results of the net primary productivity as calculated from the standing crop of the biomass at the 15th day interval are summarized in Tables 5.3 and 5.4 and

Table 5.2: Standing crop (g/plant), on the dry weight basis of field grown *Brassica juncea PR-15* as affected by UV-B radiation, individually and in combination of IAA, Kn and GA_3.

Treat-ments	*Para-meters*	*Crop Age in Days*								
		15	*30*	*45*	*60*	*75*	*90*	*105*	*120*	*135*
A	Root	0.033	0.111	0.114	0.234	0.310	0.314	0.414	0.511	0.922
	Stem	0.034	0.100	0.135	0.254	0.315	0.513	0.412	0.514	0.544
	Leaf	0.011	0.052	0.076	0.141	0.116	0.163	0.211	0.213	0.243
	Flower	–	–	–	0.035	0.048	0.053	0.063	0.065	0.083
	Fruit	–	–	–	–	0.032	0.042	0.143	0.242	0.342
	Total	**0.078**	**0.263**	**0.325**	**0.664**	**0.821**	**1.085**	**1.243**	**1.545**	**2.134**
B	Root	0.021	0.043	0.095	0.113	0.234	0.271	0.214	0.313	0.722
	Stem	0.024	0.056	0.114	0.214	0.243	0.311	0.312	0.332	0.403
	Leaf	0.046	0.032	0.053	0.112	0.095	0.132	0.114	0.142	0.182
	Flower	–	–	–	0.011	0.039	0.039	0.049	0.052	0.063
	Fruit	–	–	–	–	0.015	0.035	0.045	0.141	0.245
	Total	**0.091**	**0.0131**	**0.262**	**0.450**	**0.626**	**0.788**	**0.734**	**0.980**	**1.615**
C	Root	0.030	0.053	0.104	0.213	0.290	0.291	0.351	0.451	0.823
	Stem	0.031	0.075	0.115	0.224	0.286	0.351	0.341	0.355	0.505
	Leaf	0.082	0.045	0.061	0.121	0.105	0.141	0.189	0.192	0.214
	Flower	–	–	–	0.021	0.042	0.044	0.052	0.053	0.071
	Fruit	–	–	–	–	0.020	0.039	0.076	0.192	0.299
	Total	**0.143**	**0.173**	**0.280**	**0.579**	**0.743**	**0.866**	**1.009**	**1.243**	**1.912**
D	Root	0.031	0.063	0.103	0.213	0.281	0.292	0.352	0.421	0.822
	Stem	0.032	0.074	0.104	0.225	0.274	0.351	0.341	0.343	0.525
	Leaf	0.072	0.048	0.071	0.121	0.105	0.142	0.183	0.185	0.214
	Flower	–	–	–	0.022	0.042	0.043	0.053	0.053	0.072
	Fruit	–	–	–	–	0.021	0.039	0.076	0.186	0.299
	Total	**0.135**	**0.185**	**0.278**	**0.581**	**0.723**	**0.867**	**1.005**	**1.188**	**1.932**
E	Root	0.032	0.053	0.103	0.213	0.291	0.292	0.341	0.461	0.824
	Stem	0.032	0.075	0.124	0.224	0.291	0.361	0.334	0.357	0.505
	Leaf	0.084	0.046	0.071	0.121	0.105	0.142	0.194	0.191	0.214
	Flower	–	–	–	0.022	0.043	0.043	0.053	0.053	0.072
	Fruit	–	–	–	–	0.024	0.039	0.078	0.194	0.298
	Total	**0.148**	**0.164**	**0.298**	**0.580**	**0.754**	**0.877**	**1.00**	**1.256**	**1.913**

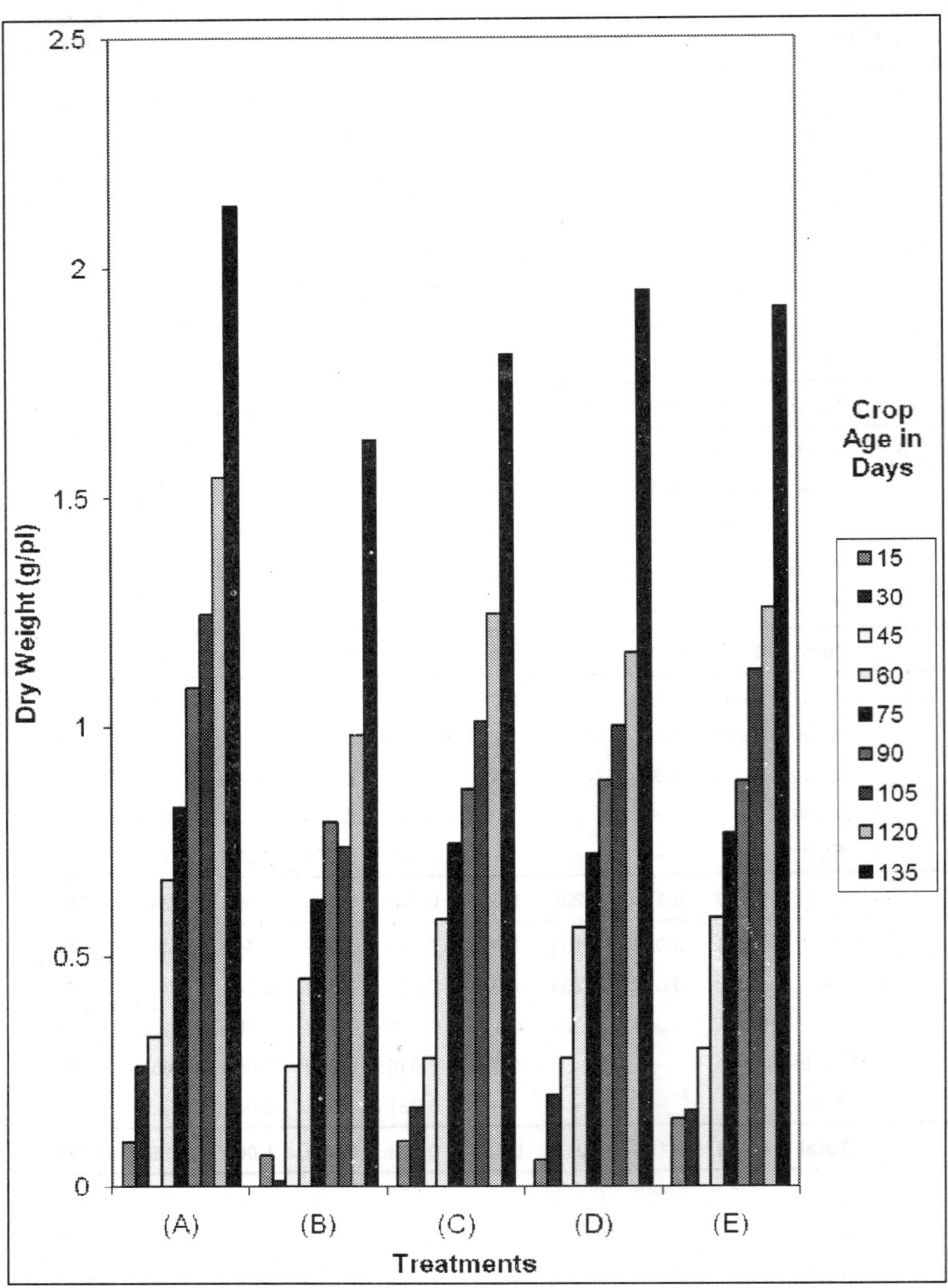

Figure 5.2: Total standing crop (g/plant), on the dry weight basis of field *Brassica juncea PR-15* as affected by different treatments.

Figures 5.3 and 5.4. The whole result followed the trend of the standing crop of biomass and indicates that the total net production of both the varieties of mustard crops were affected due to UV-B radiation (3 hrs. daily), alone and along with different plant growth regulators (PGRs). Generally, UV-B exposure was inhibited the total net production of the different plant parts *viz.* (Root + stem + Leaf + flower + Fruits). But promotory affect was reported when UV-B exposure was given along with different plant growth regulators (PGRs).

Table 5.3, demonstrated that at the 15th day stage of the growth, the values of net primary productivity and total net productivity of the *Brassica compestris PT-303* (Brown sarson) was recorded to 0.035, 0.034, 0.084 and 0.0153 g/pl in the control plot. T.N.P was observed to increase continuously up to maturity and recorded as 2.098 respectively. When the crop was exposed to UV-B radiation, the maximum inhibition of T.N.P was recorded at the 105th day stage of the growth. When the crop was sprayed along with different plant growth regulators (PGRs), such as IAA (10^{-7} M) concentration with UV-B radiation, the maximum value of T.N.P were recorded at the 105th day stage of growth and increased by ca. 17.9 per cent as compared to the individual treatment of UV-B radiation. When the crop was exposed by UV-B irradiation along with Kn (10^{-5} M) concentration, the maximum value of T.N.P. was found at 135th day stage of growth and increased by ca. 25.2 per cent, as compared to the UV-B exposure only. When the crop was exposed to UV-B radiation along with GA_3 (10^{-7} M) concentration, the promotory affects of T.N.P were recorded maximum at the 135th day stage of growth and increased by ca. 26.7 per cent as compared to the individual treatment of UV-B radiation (3 hrs. daily).

Brassica juncea PR-15

Table 5.4 exhibits that at the 15th day stage of the growth the value of the total net productivity of the *Brassica juncea PR-15* was recorded to be 0.033, 0.035, 0.110 and 0.178 g/pl in the control plot and T.N.P was observed to increase continuously up to maturity and observed as 2.135 g/pl. The Plot B, showed the inhibition in terms of the T.N.P. The maximum inhibition was reported at the maturity stage of growth and decreased by ca. 24.6 per cent as compared to the control. When the crop was exposed by UV-B irradiation along with IAA (10^{-7} M), the maximum value of T. N. P. was observed at the 75th day interval and increased by ca. 12.7 per cent as compared to UV-B exposed plot. When the crop was exposed by UV-B radiation along with Kn (10^{-5} M), the maximum value of T.N.P was recorded at the maturity and increased by ca. 25.2 per cent as compared to UV-B exposure only. When the crop was exposed to UV-B radiation along with GA_3 (10^{-7} M), the maximum promotory effect was also recorded at the maturity stage and increased by ca. 25.3 per cent as compared to the UV-B exposure only (3 hrs. daily).

Compartmental Partitioning of Biomass

The fluctuation in the total biomass is often accompanied by the substantial change in the partitioning of the biomass into plant organs. The data of the percent contribution of the biomass in various parts of the plants are set in the Tables 5.5 and 5.6, for the both varieties of mustard crops. In case of the *Brassica compestris PT-303*,

Table 5.3: Net primary productivity on dry weight basis (g/plant/15 days) of root, stem, leaf, flower and fruit of field grown *Brassica compestris PT-303* as affected by UV-B radiation, individually and in combination of IAA, Kn and GA_3.

Treat-ments	*Para-meters*	*Crop Age in Days*								
		15	*30*	*45*	*60*	*75*	*90*	*105*	*120*	*135*
A	Root	0.035	0.091	0.109	0.234	0.310	0.314	0.413	0.511	0.822
	Stem	0.034	0.102	0.134	0.315	0.414	0.513	0.512	0.514	0.544
	Leaf	0.084	0.091	0.106	0.112	0.156	0.212	0.214	0.311	0.315
	Flower	–	–	–	0.034	0.048	0.53	0.058	0.062	0.073
	Fruit	–	–	–	–	0.034	0.043	0.145	0.242	0.342
	Total	**0.153**	**0.284**	**0.349**	**0.695**	**0.962**	**1.135**	**1.342**	**1.641**	**2.098**
B	Root	0.030	0.043	0.095	0.098	0.232	0.271	0.214	0.312	0.622
	Stem	0.024	0.055	0.104	0.254	0.244	0.313	0.320	0.332	0.323
	Leaf	0.052	0.054	0.074	0.142	0.114	0.142	0.142	0.213	0.235
	Flower	–	–	–	0.013	0.036	0.039	0.044	0.053	0.066
	Fruit	–	–	–	–	0.019	0.035	0.045	0.121	0.245
	Total	**0.106**	**0.152**	**0.273**	**0.507**	**0.645**	**0.800**	**0.765**	**1.031**	**1.491**
C	Root	0.032	0.053	0.103	0.223	0.280	0.301	0.321	0.412	0.724
	Stem	0.032	0.057	0.124	0.264	0.276	0.341	0.341	0.354	0.494
	Leaf	0.64	0.069	0.081	0.152	0.125	0.151	0.164	0.251	0.259
	Flower	–	–	–	0.022	0.029	0.039	0.052	0.059	0.069
	Fruit	–	–	–	–	0.021	0.040	0.068	0.142	0.293
	Total	**0.128**	**0.179**	**0.308**	**0.661**	**0.731**	**0.872**	**0.946**	**1.218**	**1.836**
D	Root	0.030	0.052	0.102	0.213	0.291	0.302	0.322	0.411	0.723
	Stem	0.030	0.55	0.124	0.265	0.274	0.312	0.341	0.352	0.524
	Leaf	0.064	0.058	0.082	0.152	0.125	0.151	0.174	0.253	0.257
	Flower	–	–	–	0.02	0.044	0.047	0.043	0.054	0.065
	Fruit	–	–	–	–	0.029	0.038	0.068	0.144	0.283
	Total	**0.124**	**0.165**	**0.308**	**0.652**	**0.763**	**0.850**	**0.948**	**1.216**	**1.852**
E	Root	0.032	0.053	0.103	0.223	0.291	0.298	0.331	0.420	0.724
	Stem	0.025	0.056	0.124	0.264	0.281	0.321	0.351	0.362	0.525
	Leaf	0.064	0.059	0.081	0.153	0.126	0.151	0.174	0.262	0.286
	Flower	–	–	–	**0.022**	**0.044**	**0.048**	**0.053**	**0.055**	**0.064**
	Fruit	–	–	–	–	**0.029**	**0.039**	**0.073**	**0.146**	**0.254**
	Total	**0.121**	**0.168**	**0.308**	**0.662**	**0.771**	**0.857**	**0.982**	**1.245**	**1.853**

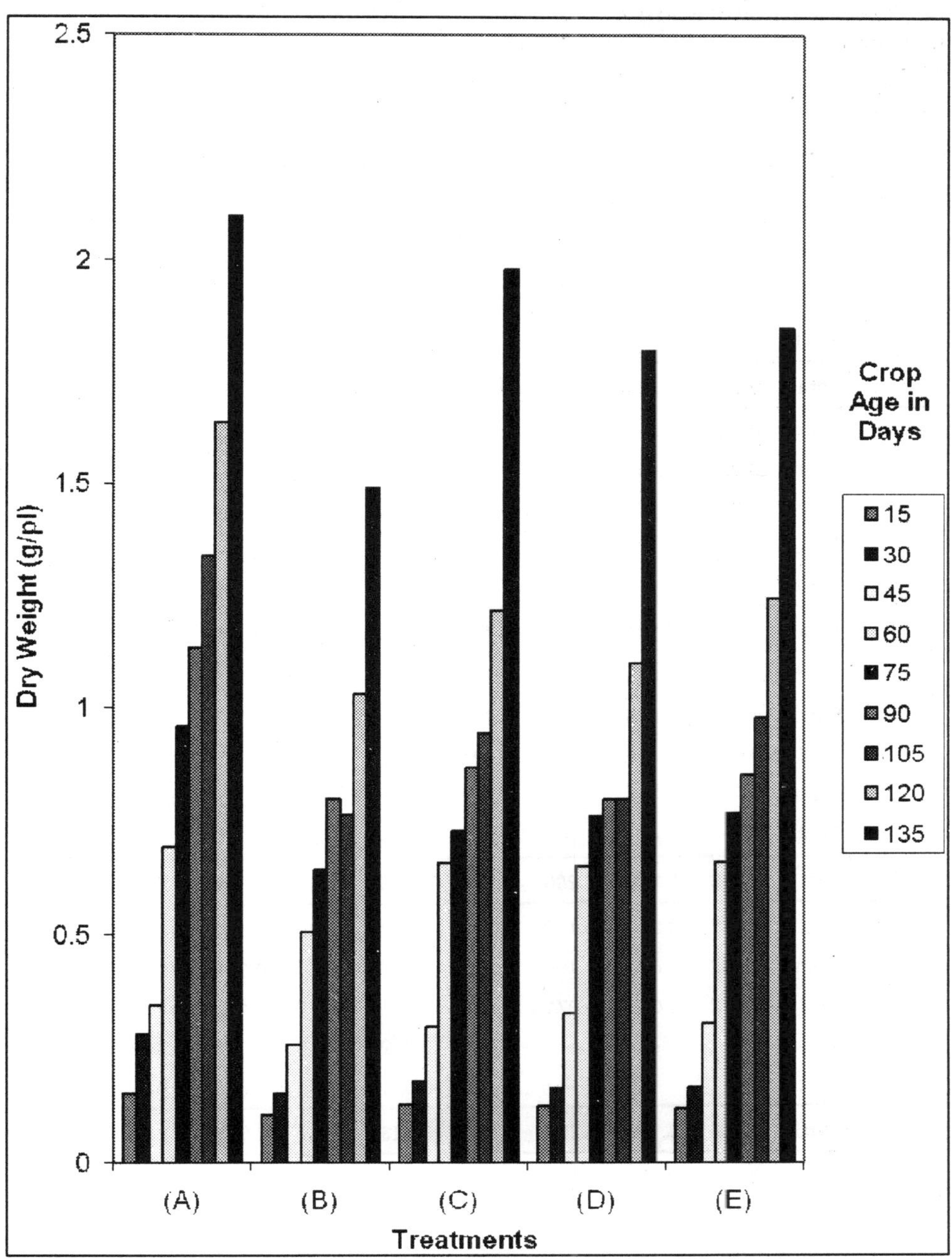

Figure 5.3: Total net primary productivity on the dry weight basis (g/plant/15 days) of *Brassica compestris PT-303* as affected by different treatments.

Table 5.4: Net primary productivity on dry weight basis (g/plant/15 days) of root, stem leaf, flower and fruit of field grown *Brassica juncea PR-15* as affected by UV-B radiation, individually and in combination of IAA, Kn and GA_3.

Treat-ments	*Para-meters*	*Crop Age in Days*								
		15	*30*	*45*	*60*	*75*	*90*	*105*	*120*	*135*
	Root	0.033	0.111	0.114	0.234	0.310	0.314	0.414	0.511	0.922
	Stem	0.034	0.100	0.135	0.254	0.315	0.513	0.412	0.514	0.544
A	Leaf	0.011	0.052	0.076	0.141	0.116	0.163	0.211	0.213	0.243
	Flower	–	–	–	0.035	0.048	0.053	0.063	0.065	0.083
	Fruit	–	–	–	–	0.032	0.042	0.143	0.242	0.342
	Total	**0.078**	**0.263**	**0.325**	**0.664**	**0.821**	**1.085**	**1.243**	**1.545**	**2.134**
	Root	0.021	0.043	0.095	0.113	0.234	0.271	0.214	0.313	0.722
	Stem	0.024	0.056	0.114	0.214	0.243	0.311	0.312	0.332	0.403
B	Leaf	0.046	0.032	0.053	0.112	0.095	0.132	0.114	0.142	0.182
	Flower	–	–	–	0.011	0.039	0.039	0.049	0.052	0.063
	Fruit	–	–	–	–	0.015	0.035	0.045	0.141	0.245
	Total	**0.091**	**0.131**	**0.262**	**0.450**	**0.626**	**0.788**	**0.734**	**0.980**	**1.615**
	Root	0.030	0.053	0.104	0.213	0.290	0.291	0.351	0.451	0.823
	Stem	0.031	0.075	0.115	0.224	0.286	0.351	0.341	0.355	0.505
C	Leaf	0.082	0.045	0.061	0.121	0.105	0.141	0.189	0.192	0.214
	Flower	–	–	–	0.021	0.042	0.044	0.052	0.053	0.071
	Fruit	–	–	–	–	0.020	0.039	0.076	0.192	0.299
	Total	**0.143**	**0.173**	**0.280**	**0.579**	**0.743**	**0.866**	**1.009**	**1.243**	**1.912**
	Root	0.031	0.063	0.103	0.213	0.281	0.292	0.352	0.421	0.822
	Stem	0.032	0.074	0.104	0.225	0.274	0.351	0.341	0.343	0.525
D	Leaf	0.072	0.048	0.071	0.121	0.105	0.142	0.183	0.185	0.214
	Flower	–	–	–	0.022	0.042	0.043	0.053	0.053	0.072
	Fruit	–	–	–	–	0.021	0.039	0.076	0.186	0.299
	Total	**0.135**	**0.185**	**0.278**	**0.581**	**0.723**	**0.867**	**1.005**	**1.188**	**1.932**
	Root	0.032	0.053	0.103	0.213	0.291	0.292	0.341	0.461	0.824
	Stem	0.032	0.075	0.124	0.224	0.291	0.361	0.334	0.357	0.505
E	Leaf	0.084	0.046	0.071	0.121	0.105	0.142	0.194	0.191	0.214
	Flower	–	–	–	0.022	0.043	0.043	0.053	0.053	0.072
	Fruit	–	–	–	–	0.024	0.039	0.078	0.194	0.298
	Total	**0.148**	**0.164**	**0.298**	**0.580**	**0.754**	**0.877**	**1.00**	**1.256**	**1.913**

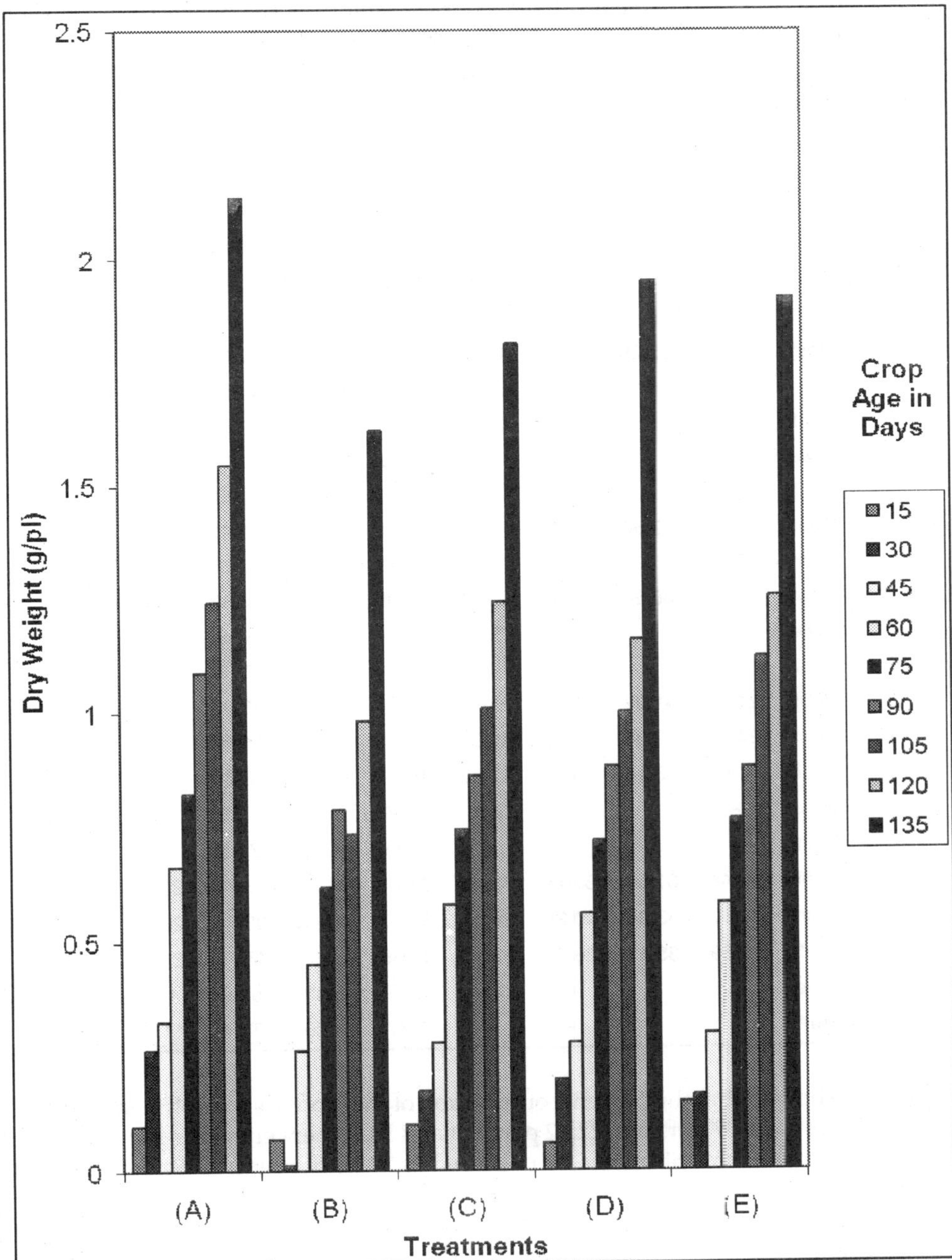

Figure 5.4: Total net primary productivity on the dry weight basis (g/plant/15 days) of *Brassica juncea PR-15* as affected by different treatments.

Table 5.5: Percentage contribution of root, stem, leaf, flower to the total standing crop of field grown *Brassica compestris PT-303* (Brown sarson) as affected by UV-B radiation (3 hrs. daily), individually and in combination of IAA, Kn and GA_3.

Treat-ments	*Para-meters*	*Crop Age in Days*								
		15	*30*	*45*	*60*	*75*	*90*	*105*	*120*	*135*
A	Root	22.87	32.04	31.23	24.24	32.26	27.68	30.77	31.13	39.18
	Stem	22.22	35.91	38.39	45.32	43.03	45.19	38.15	31.32	25.92
	Leaf	54.90	32.02	30.37	16.11	16.23	18.67	15.95	18.95	15.06
	Flower	–	–	–	4.89	4.98	4.66	4.34	3.78	3.47
	Fruit	–	–	–	–	3.53	3.79	10.83	14.74	16.34
B	Root	28.30	28.28	34.79	19.32	35.96	33.87	27.97	30.27	41.71
	Stem	22.64	36.18	38.09	50.09	37.82	39.12	41.83	32.20	21.66
	Leaf	49.05	35.52	27.10	28.00	17.67	17.75	18.56	20.65	15.76
	Flower	–	–	–	2.56	5.58	48.75	5.75	5.14	4.42
	Fruit	–	–	–	–	29.45	43.75	5.89	11.73	16.43
C	Root	25.00	29.60	33.44	33.73	38.30	34.51	33.93	33.82	39.43
	Stem	25.00	31.94	40.25	39.93	37.75	39.10	36.04	29.06	26.90
	Leaf	50.00	38.54	26.29	22.99	17.09	17.31	17.33	20.60	14.11
	Flower	–	–	–	3.32	3.96	3.32	5.49	4.85	3.76
	Fruit	–	–	–	–	2.87	3.49	7.19	11.65	15.96
D	Root	24.19	41.93	33.12	32.67	38.14	35.93	33.97	33.79	39.04
	Stem	24.19	33.34	40.25	40.65	35.91	36.70	35.97	28.95	28.30
	Leaf	51.61	35.16	26.27	23.32	16.39	17.77	18.36	20.83	13.87
	Flower	–	–	–	3.38	5.76	5.52	4.56	4.45	3.53
	Fruit	–	–	–	–	3.83	4.48	7.18	11.84	15.28
E	Root	26.44	31.54	33.45	33.69	37.74	34.78	33.70	33.73	39.08
	Stem	20.67	33.34	40.25	39.87	36.47	37.46	35.75	29.08	28.34
	Leaf	52.89	35.12	26.30	23.12	16.35	17.61	17.72	21.04	15.44
	Flower	–	–	–	3.34	5.73	5.63	5.39	4.43	3.46
	Fruit	–	–	–	–	3.77	4.56	7.44	11.73	13.70

the results showed that the percent contribution of the biomass of root, stem and leaf was observes as 22.87 per cent, 22.22 per cent and 54.90 per cent in the plot-A. In the plot-B, the percent contribution was observed as 28.30 per cent, 22.64 per cent and 49.05 per cent at the 15th day interval of the crop growth. When plot-C was studied at the15[th] day interval of crop growth, the data observed as ca. 25.00 per cent, 25.00 per cent, 50.00; plot-D was studied at the 15th day interval of crop growth, the data observed as ca. 24.19 per cent, 24.19 per cent, 51.61 per cent respectively. When plot-E was also studied at the15[th] day interval of crop growth, the data recorded as ca. 26.44 per cent, 20.67 per cent and 52.89 per cent respectively. When the percent contribution of the biomass of various plant part was studied at the maturity stage of crop growth in the plot-A, observed as ca. 39.18 per cent, 25.52 per cent, 15.06 per cent, 3.47 per cent and 16.34 per cent for the root, stem, leaf, flower and fruit respectively.

The Table 5.6 represents the percent contribution of the *Brassica juncea PR-15.* The percent contribution of the root, leaf and stem at the 15th day interval of crop growth was recorded as ca. 18.17 per cent, 19.42 per cent and 6.28 per cent in the control plot (A). In plot-B (UV-B), the percent contribution of root, stem, leaf as 23.07 per cent, 26.37 per cent and 50.54 per cent; in the (plot-C), the percent contribution recorded as ca. 22.97 per cent, 23.70 per cent, 53.34 per cent; in the plot (D), the percent

Table 5.6: Percentage contribution of root, stem, leaf, flower, fruit to the total standing crop of field Grown *Brassica juncea PR-15* (Rai) as affected by UV-B radiation (3 hrs. daily), individually and in combination of IAA, Kn and GA_3.

Treat-ments	*Para-meters*	*Crop Age in Days*								
		15	*30*	*45*	*60*	*75*	*90*	*105*	*120*	*135*
A	Root	18.17	42.0	35.07	35.24	37.75	28.95	33.32	33.07	43.16
	Stem	19.42	38.03	41.53	38.25	38.36	47.28	33.12	33.26	25.49
	Leaf	6.28	19.77	23.28	21.23	14.12	15.02	17.04	18.84	11.38
	Flower	---	–	–	5.27	5.84	4.89	5.07	4.20	3.89
	Fruit	–	–	–	–	3.97	3.87	11.50	15.65	16.02
B	Root	23.07	32.82	36.25	25.11	37.38	34.39	29.16	31.93	44.46
	Stem	26.37	42.74	43.51	47.55	38.81	39.46	42.51	33.88	24.82
	Leaf	50.54	24.42	20.22	24.88	15.18	16.76	15.54	14.49	11.21
	Flower	–	–	–	2.45	6.24	4.94	6.68	5.34	3.89
	Fruit	–	–	–	–	2.30	4.45	6.14	14.39	15.09
C	Root	22.97	30.64	37.14	36.79	39.03	33.61	34.78	36.29	43.04
	Stem	23.70	40.00	41.08	38.69	38.49	40.53	33.79	28.56	26.41
	Leaf	53.34	25.95	21.17	20.89	14.13	16.38	18.73	15.44	11.20
	Flower	–	–	–	3.63	5.67	4.96	5.15	4.27	3.72
	Fruit	–	–	–	–	2.69	4.49	7.14	4.27	15.64
D	Root	22.97	34.05	37.05	36.66	38.86	33.67	35.03	35.43	42.54
	Stem	23.70	40.00	37.41	38.72	37.89	40.48	33.93	28.87	27.17
	Leaf	53.34	25.95	25.53	20.82	14.53	16.38	18.20	15.57	11.08
	Flower	---	---	---	3.79	5.80	4.96	5.28	4.47	3.73
	Fruit	---	---	---	---	2.90	4.49	7.57	15.65	15.47
E	Root	21.63	32.31	34.56	36.66	38.59	33.30	34.10	36.70	43.05
	Stem	21.62	45.73	41.62	38.55	38.59	41.16	33.40	28.42	26.38
	Leaf	56.75	28.04	23.82	20.82	13.92	16.19	19.40	15.20	11.19
	Flower	–	---	---	3.78	5.70	4.90	5.30	4.22	3.77
	Fruit	---	---	---	---	3.19	4.45	7.80	15.45	15.57

contribution observed as 22.97 per cent, 23.70 per cent, 53.34 per cent and in plot (E), the percent contribution was also observed as ca. 21.63 per cent, 21.62 per cent, 56.75 per cent respectively. The percent contribution of the root, stem, leaf, flower and fruit was recorded at the maturity stage of crop. The plot (A) was showed as ca. 43.16 per cent, 25.49 per cent, 11.38 per cent, 3.89 per cent and 16.02 per cent respectively. The data was recorded to the plot (B) as ca. 44.46 per cent, 24.82 per cent, 11.21, 3.89 per cent and 15.09 per cent; in the plot (C), the data was recorded as ca. 43.04 per cent, 26.41 per cent, 11.20 per cent, 3.72 per cent and 15.64 per cent respectively. The data was noted from the plots (D) and (E), as ca. 42.54 per cent, 27.17 per cent, 11.08 per cent, 3.73 per cent, 15.47 per cent, 43.05 per cent and as ca. 26.38 per cent, 11.19 per cent, 3.77 per cent and 15.57 per cent respectively.

Storage of Dry Matter

The storage of the dry matter was showed in the Tables 5.7 and 5.8 of various plant parts of the *Brassica compestris PT-303* and *Brassica juncea PR-15* as affected by the various treatments taken into account for the study. The table 5.7 exhibits that the live vegetation stored 2.106 g/pl dry matter during the complete life span in the control. Out of these, 1.284 g/pl contributed to the above ground and 0.822 g/pl to underground plant parts. The stem accounted as 0.554 g/pl, leaves 0.315 g/pl, 0.073 g/pl flower and fruit 0.342 per cent g/pl respectively in the control plot of the *Brassica compestris PT-303*.

Table 5.7: Storage of dry matter of different plant parts of field grown *Brassica compestris PT-303* as affected by UV-B radiation (3 hrs. days), individually and in combination of IAA, Kn and GA_3.

Brassica compestris PT-303	*A*	*B*	*C*	*D*	*E*
Live Vegetation	2.106	1.490	1.839	1.852	1.853
Above Ground	1.284	0.868	1.115	1.129	1.129
Under Ground	0.822	0.622	0.724	0.723	0.724
Stem	0.554	0.322	0.494	0.524	0.525
Leaves	0.315	0.235	0.259	0.257	0.286
Flower	0.073	0.066	0.069	0.065	0.064
Fruits	0.342	0.245	0.293	0.283	0.254

When the crop was exposed to UV-B radiation only, a decrease of ca. 30 per cent was recorded in storage of the live matter of the *Brassica compestris PT-303*. The above ground parts of the dry matter were decreased by ca. 33 per cent and undergrounds parts of the dry matter were decreased by ca. 25 per cent respectively. The different above ground plant parts as stem, leaves, flower and fruits etc. were decreased by ca. 42 per cent, 26 per cent, 10 per cent and 29 per cent respectively as compared to control.

When the crop was exposed by UV-B along with IAA (10^{-7} M), the data observed an increased by ca. 23 per cent, 28 per cent, 16.3 per cent, 53 per cent, 10.5 per cent, 10.6 per cent and 19.9 per cent respectively for the live vegetation, above ground, under ground, stem, leaves, flower and fruits respectively. When UV-B exposure was given along with Kn (10^{-5} M) the data was observed an increased by ca. 24.3 per cent, 30. per cent, 16.5 per cent, 62 per cent, 10 per cent 5.5 per cent and 16.5 per cent for the live vegetation, above ground, under ground, stem, leaves, flowers and fruit respectively. When UV-B exposure was given along with GA_3 (10^{-6} M), the data showed an increase of ca. 24.5 per cent, 30 per cent, 16.5 per cent, 63 per cent, 21.7 per cent, 4 per cent and 3.6 per cent respectively for the live vegetation above ground, under ground, stem leaves, flower and fruits as compared to the individual treatment of UV-B radiation.

Perusal of the data in Table 5.8 indicates that live vegetation stored 2.134 g/pl dry matter during the whole life span of *Brassica juncea PR-15* in the control plot. Out of these 1.212 g/pl contributed to the above ground and 0.922 g/pl contributed to underground parts of plant as stem accounted for 0.544 g/pl, leaves 0.243 g/pl, flower 0.083 g/pl and fruits 0.342 g/pl respectively. When the crop was treated with UV-B only, it resulted into a decrease of ca. 25 per cent in the storage of live matter of *Brassica juncea PR-15.* The above ground dry matter was decreased by ca. 27 per cent and under ground decreased by ca. 22 per cent respectively. The various above ground plant parts *viz.* stem, leaves, flower and fruits etc. resulted into a reduction and reduced by ca. 26.7 per cent, 26.5 per cent, 25.9 per cent and 29.6 per cent respectively, as compared to the control plot.

Table 5.8: Storage of dry matter of different plant parts of field grown *Brassica juncea PR-15* as affected by UV-B radiation affected by UV-B radiation (3-hrs. daily), individually and in combination of IAA, Kn and GA_3.

Brassica juncea PR-15	*A*	*B*	*C*	*D*	*E*
Live Vegetation	2.134	1.615	1.912	1.932	1.913
Above Ground	1.212	0.893	1.089	1.110	1.089
Under Ground	0.922	0.722	0.823	0.822	0.824
Stem	0.544	0.403	0.505	0.525	0.505
Leaves	0.243	0.182	0.214	0.214	0.214
Flower	0.083	0.063	0.071	0.072	0.073
Fruits	0.342	0.245	0.299	0.298	0.297

When the crop was subjected to combined treatments of UV-B + IAA (10^{-7}M), the data showed an increase by ca.18.9 per cent, 21.9 per cent, 13.8 per cent, 25.3 per cent 17.8 per cent, 12.9 per cent and 22.4 per cent for the live vegetation, above ground, underground, stem, leaves, flower and fruits respectively. When the UV-B exposure was given along with UV-B + Kn (10^{-5}M), the data was reported and increased by ca. 19.7 per cent, 24.4 per cent, 13.8 per cent, 30.3 per cent, 17.6 per cent, 14.9 per cent and 21.7 per cent for the live matter, above ground, under ground, stem, leaves, flower and fruit respectively. In case of UV-B + GA_3 (10^{-7}M), the data was showed and increased by ca. 18.6 per cent, 21.9 per cent, 14.2 per cent, 25.4 per cent, 17.6 per cent, 15.9 per cent and 21.4 per cent respectively for live vegetation, under ground, above ground, stem, leaves, flower and fruits as compared to individual UV-B exposure.

Dry Matter Dynamics

Tables 5.9 and 5.10, showed the dynamics of dry matter of *Brassica compestris PT-303* and *Brassica juncea PR-15* as affected by the individual treatment of UV-B, along with different concentration of some plant growth regulators (PGRs), such as IAA (10^{-7}M), Kn (10^{-5}M) and GA_3 (10^{-6}) in *B. compestris PT-303* and (10^{-7}M) in *B. juncea PR-15* respectively. An analysis of data of table 5.9 indicates that the total net production was amounted to 2.098, 1.491, 1.836, 1.452 and 1.853 g/pl in the plot A, B, C, D, and E respectively. The transfer of total net production to above ground was accounted 0.822 g/pl and to under ground 1.276 g/pl in the plot A (control). The stem, leaf, flower and fruit received as 1.554, 1.783, 2.025 and 1.756 g/pl of the dry matter production from T.N.P. respectively in the control plot (A) of the *Brassica compestris PT-303*.

Table 5.9: Dry matter dynamics (g/pl) of different parts of field grown *Brassica compestris PT-303* as affected by UV-B radiation (3 hrs. daily), individually and in combination of IAA, Kn and GA_3.

Treatments	*A*	*B*	*C*	*D*	*E*
TNP	2.098	1.491	1.836	1.852	1.853
TNP to ANP	0.822	0.622	0.724	0.723	0.724
TNP to UNP	1.276	0.869	1.112	1.129	1.128
TNP to Root	1.276	0.869	1.112	1.129	1.128
TNP to Stem	1.554	1.168	1.342	1.328	1.328
TNP to Leaf	1.783	1.256	1.577	1.595	1.567
TNP to Flower	2.025	1.425	1.767	1.787	1.789
TNP to Fruit	1.756	1.207	1.500	1.530	1.535

When crop subjected to UV-B (3 hrs. daily) only, a decrease of ca. 29 per cent of dry matter production was recorded in the T.N.P. The decrease in the dry matter which was transferred from the T.N.P to above ground N.P., underground N.P., stem, leaf, flower and fruit were calculated as to ca. 25 per cent, 32 per cent, 25.2 per cent, 30.4 per cent, 30.3 per cent and 32.7 per cent respectively, as compared to individual treatment of UB-B exposure. When the crop exposed to UV-B radiation, along with IAA (10^{-7}M) concentration, the maximum mitigation was observed over UV-B damage. The T.N.P to upper ground level was noticed to increased by ca. 16.4 per cent; to underground level ca. 27.9 per cent; to stem ca. 14.9 per cent; to leaf ca. 25.6 per cent; to flower ca. 24 per cent and to fruit ca. 24.3 per cent respectively as compared to the UV-B exposure only.

When the crop was exposed by UV-B radiation along with Kn (10^{-5} M), the maximum mitigation of T.N.P was observed and noticed as above ground level to increased by ca.16.3 per cent; under ground level to ca. 29.9 per cent; to stem ca.13.7 per cent; to leaves ca. 26.9 per cent; to flower ca. 25.5 per cent and to fruit ca. 26.7 per cent respectively as compared to the over UV-B exposure. When the crop was also subjected to combined treatment of UV-B along with GA_3 (10^{-7}M) concentration, the maximum mitigation of T.N.P was recorded as above ground to increased by ca. 16.4 per cent; to underground by ca. 29.8 per cent; to stem ca. 13.7 per cent; to leave ca. 24.8 per cent; to flower ca. 25.6 per cent and fruit ca. 27.2 per cent respectively as compared to individual treatment of UV-B radiation.

A perusal of data in Table 5.10 indicate that the total net productivity in *Brassica juncea PR-15* was amounted to ca. 2.134, 1.624, 1.912, 1.931 and 1.91 g/pl in the plot A, B. C, D and E respectively. The transfer of total net production to above ground was amounted 0.622 g/pl; to underground 1.212 g/plant; to stem 1.590 g/pl; to leaves 1.891 g/pl; to flower 2.051 g/pl and to fruits 1.792 g/pl respectively of the dry matter production from the T.N.P in control plot.

Table 5.10: Dry matter dynamics (g/pl) of different parts of field grown *Brassica juncea PR-15* as affected by UV-B radiation (3 hrs. daily), individually and in combination of IAA, Kn and GA_3.

Parameters	A	B	C	D	E
TNP	2.134	1.624	1.912	1.931	1.914
TNP to ANP	0.622	0.722	0.823	0.822	0.824
TNP to UNP	1.212	0.902	1.089	1.109	1.090
TNP to Root	1.212	0.902	1.089	1.109	1.090
TNP to Stem	1.590	1.221	1.407	1.406	1.406
TNP to Leaf	1.891	1.442	1.698	1.717	1.700
TNP to Flower	2.051	1.561	1.841	1.859	1.841
TNP to Fruit	1.792	1.315	1.580	1.600	1.569

A decrease in dry matter which is transferred from the T.N.P to ANP, UNP, stem, leaf, flower and fruit was observed as 22.5, 24.3, 24.4, 24.3, 24.2, 25.0 and 27.2 g/plant respectively, when the crop was exposed to UV-B only. When the crop was exposed to UV-B radiation along with IAA (10^{-7} M), the transfer of T.N.P to above ground was noticed to increased by ca. 14 per cent; to under ground level ca. 20.7 per cent; to stem ca. 16.3 per cent; to leaf ca. 17.7 per cent; to flower ca. 18.9 per cent and to fruit ca. 20.3 per cent respectively as compared to individual treatment of UV-B radiation. When the crop was exposed to UV-B radiation along with Kn (10^{-5}M), the transfer of T.N.P to above ground was observed to increased by ca. 13.4 per cent; to under ground 22.9 per cent; to stem 16 per cent; to leaf 19.5 per cent; to flower 20 per cent and to fruit 21.7 per cent respectively as compared to individual treatment of UV-B radiation. When the crop was also exposed to UV-B radiation along with GA_3 (10^{-7} M), the transfer of T.N.P to above ground was noticed to increased by ca. 14.6 per cent; to under ground 20.9 per cent; to stem 15.6 per cent; to leaf 17.9 per cent; to flower 18.4 per cent and to fruits 19.4 per cent respectively as compared to individual treatment of UV-B radiation.

Of fundamental importance to the economic yield is the distribution of the dry matter in the storage and vegetative parts. By all the treatments, the crop Harvesting Index (HI) was observed to be affected significantly. When individual treatment of UV-B exposure was given to crops, it has been noticed that harvesting index was reduced. The Table 5.11 showed that HI of *Brassica compestris PT-303* was amounted 0.83, 0.80, 0.81, 0.82 and 0.82 in the plots as A, B, C, D and E respectively.

Table 5.11: Yield attributes of field grown *Brassica compestris PT-303* as affected by UV-B radiation (3 hrs. daily), individually and in combination of IAA, Kn and GA_3.

Treatments	*Biomass (g/plant)*		*Net Primary Productivity*	*Harvesting Index*	*Shelling Percentage*
	Fruit	*Seed*			
A	1.756	1.655	2.098	0.83	94.24
B	1.207	1.130	1.491	0.80	90.69
C	1.500	1.448	1.836	0.81	93.73
D	1.530	1.459	1.852	0.82	92.71
E	1.535	1.457	1.853	0.82	91.11

Table 5.12 of *Brassica juncea PR-15,* showed that HI was amounted 0.83, 0.80, 0.82, 0.82 and 0.81 in the plot A, B, C, D and E respectively. When plot-B was exposed to UV-B radiation only, it was reported the marked reduction in HI and decreased by ca. 4 per cent as compared to control plot (A). When plants were sprayed with plant growth regulators (PGRs), along with UV-B exposure, exhibited a general promotion of the Harvesting Index (HI) as compared to UV-B radiation. The Kn (10^{-5}M) and GA_3 (10^{-6}M) concentrations were noticed to be most effective in promoting to HI in *Brassica compestris PT-303* and promoted ca. 3 per cent and 2.5 per cent increased as compared to individual treatment of UV-B radiation.

Table 5.12: Yield attributes of field grown *Brassica juncea PR-15* as affected by UV-B radiation (3 hrs. daily), individually and in combination of IAA, Kn and GA_3.

Treatments	*Biomass (g/plant)*		*Net Primary Productivity*	*Harvesting Index*	*Shelling Percentage*
	Fruit	*Seed*			
A	1.792	1.657	2.134	0.83	92.46
B	1.315	1.196	1.624	0.80	90.95
C	1.580	1.449	1.912	0.82	91.70
D	1.600	1.470	1.931	0.82	91.87
E	1.569	1.440	1.914	0.81	91.77

In case of *Brassica juncea PR-15,* the crop exposed to UV-B radiation only (3 hrs. daily), it was observed deleterious to the harvesting Index and decreased by ca. 4.3 per cent as compared to control. When plant was sprayed with plant growth regulators (PGRs) along with UV-B radiation, exhibited a general promotion of the harvesting Index (HI) as compared to UV-B radiation. The IAA (10^{-7}M) and Kn (10^{-5}M) were applied and noticed to be the most effective mitigatory and promoted ca. 2.5 per cent and 2.4 per cent increased as compared to individual treatment of UV-B radiation (3 hrs. daily).

The percent seed explains the percentage of seed biomass over fruit biomass, including shell. Higher is the percent seed in a particular treatment more efficient is the treatment to improved the seed production. In the control plot (A) of *Brassica compestris PT-303* the shelling percentage was observed to ca. 94.24 per cent (Table 5.11). In UV-B exposed plot (3 hrs. daily), shelling percentage was noticed and decreased by ca. 5 per cent as compared to control. When crop was exposed to UV-B radiation, along with plant growth regulators *viz.* IAA Kn and GA_3 concentrations were effective and promotion in the shelling percentage as compared to individual treatment of UV-B radiation. Amongst plant growth regulators (PGRs), IAA was noted most effective and shelling percentage was noted to increase by ca. 5 per cent as compared to individual treatment of UV-B exposure.

Table 5.12 represents the shelling percentage in the *Brassica juncea PR-15* was recorded as ca. 92.46 per cent in the control plot. When crop was exposed to individual treatment of UV-B radiation the significant reduction was observed in the shelling percentage and decreased by ca. 3 per cent as compared to control. When UV-B exposure was given along with IAA, Kn and GA_3 (PGRs), a promotion was observed in shelling percentage as compared to UV-B exposure. The Kn was observed most counteracting against UV-B inhibition on shelling percentage and increased by ca. 6 per cent as compared to individual treatment of UV-B exposure.

Chapter 6

Deleterious Impacts of UV-B Alone and Alongwith Certain Plant Growth Regulators, on Some Physiological and Biological Aspects

Previous experiments have shown that UV-B radiation influenced the crop growth and development in terms of the plant height, leaf area, fresh weight, dry weight, harvesting index and shelling percentage. The metigatory effects of some plant growth regulators on most of these parameters were also observed during different experimental work. These effects are directly or indirectly related to the physiological processes ot the crop plants. It can also be said that these morphological changes caused by UV-B radiation, which are the result of physiological distortion of the plants. In the present investigation, so it is desired to investigate the some physiological parameters as compared to individual exposure of UV-B radiation and in combination of some plant growth regulators (PGRs) *viz.* IAA, Kn and GA_3 concentrations.

Chlorophyll Pigment during Seedling Growth

Surface sterilized seeds of both the varieties of mustard crops such as Brown sarson and Rai were imbibed in the water for 6-hrs. and then seeds were washed by distilled water and transferred to Petridihes, for the seed germination and seedling growth studies and exposed with UV-B radiation (3-hrs. daily), alone and along with various concentration of these plant growth regulators. Chlorophyll a, b and protochlorophyll were measured after 7 days of growth in both varieties of the mustard crops as explained in material and methods. The data are presented in the Table 6.1 for *Brassica compestris PT-303* and Table 6.2 for *Brassica juncea PR-15.*

Table 6.1, showed a perusal of data that chlorophyll a (mg/pl), chlorophyll b (mg/pl), protochlorophyll (mg/pl) and ratio of chlorophyll a and b in the control set were amounted as 0.45±0.05, 0.39±0.03, 0.50±0.07 and 1.15 respectively. When the germinating seedlings growth was studied with UV-B radiation only, it observed that marked decline in the contents of various chlorophyll pigments. The inhibition was observed as ca. 23 per cent, 17 per cent, 6 per cent and 8 per cent of chlorophyll a, chlorophyll b, protochlorophyll and a/b ratio respectively as compared to control. When sets C, D and E were observed with (PGRs + UV-B), a general promotion was noticed in these pigments as compared to individual treatment of UV-B radiation (set-B) only. The IAA was observed to be the most effective PGR to counteract the UV-B induced deleterious impacts for all the studied chlorophyll pigments. The result observed as ca. 14.7 per cent, 12.1 per cent, 10 per cent and 3.3 per cent increase over the UV-B treatment (set-B) in chlorophyll a, b, protochlorophyll and a/b ratio respectively.

Table 6.1: Chlorophyll content at the seedling stage after 7 days of germination as affected by UV-B radiation (3 hrs. daily), individually and in combination of IAA, Kn and GA_3 in *Brassica compestris PT-303.*

Treatments	*Chlorophyll a*	*Chlorophyll b*	*Protochlorophyll*	*a/b ratio*
A	**0.45±0.05**	**0.39±0.03**	**0.50±0.07**	1.15
B	**0.38±0.02**	**0.30±0.01**	**0.47±0.04**	0.78
C	**0.39±0.03**	**0.33±0.02**	**0.49±0.03**	0.84
D	**0.34±0.07**	**0.35±0.03**	**0.48±0.02**	0.97
E	**0.39±0.03**	**0.34±0.01**	**0.50±0.06**	1.14

An observation of result in Table 6.2, showed that in *Brassica juncea PR-15,* the chlorophyll a (mg/pl), chlorophyll b (mg/pl), protochlorophyll and ratio of the chlorophyll a/b was studied in the control set-(A) and amounted as 0.44±0.04, 0.43±0.03 0.95±0.05 and 1.02 respectively. When the seedlings were studied with UV-B radiation alone, it showed a marked decline in the contents of chlorophyll pigments except a/b ratio. The chlorophyll development was observed and inhibited by ca. 16 per cent, 24 per cent, 6 per cent and 33 per cent respectively in terms of chlorophyll a, b protochlorophyll and chlorophyll a/b ratio by UV-B treatments. When the seedlings were exposed to UV-B exposure along with different PGRs, a general promotion was recorded in all the chlorophyll pigments as compared to individual treatment of UV-B exposure. The IAA was found to record promotion as ca. 2.6 per cent, 10 per cent, 4.3 per cent and 7.7 per cent respectively in chlorophyll a, b, protochlorophyll and chlorophyll a/b ratio. Kn was found to recorded promotion as ca. 16.6 per cent for chlorophyll b, 2.2 per cent for protochlorophyll and 46 per cent for a/b ratio respectively and GA_3 showed a promotion as ca. 2.7 per cent for chlorophyll a, 13.4 per cent for chlorophyll b, 6.4 per cent for protochlorophyll and 46 per cent for a/b ratio respectively. But in case of chlorophyll a, the Kn was identified to cause inhibition by ca.11 per cent as compared to UV-B radiation alone.

Table 6.2: Chlorophyll content at seedling stage after 7 days of germination as affected by UV-B radiation (3 hrs. daily), individually and in combination of IAA, Kn and GA_3 in *Brassica juncea PR-15.*

Treatments	*Chlorophyll a*	*Chlorophyll b*	*Protochlorophyll*	*a/b ratio*
A	**0.44**±0.04	**0.43**±0.03	**0.45**±0.05	1.02
B	**0.34**±0.01	**0.36**±0.02	**0.40**±0.03	0.94
C	**0.39**±0.03	**0.40**±0.04	**0.44**±0.04	0.97
D	**0.38**±0.06	**0.35**±0.05	**0.40**±0.01	1.08
E	**0.35**±0.02	**0.32**±0.01	**0.41**±0.05	1.09

Chlorophyll Analysis during Crop Growth

Effects of UV-B exposure alone and along with some plant growth regulators on the chlorophyll development were also observed in the same two varieties of the mustard crops, which were grown earlier for the growth patterns studies. For the chlorophyll estimation, the plants were selected regularly at the 15 days interval from the seedling emergence upto maturity.

Table 6.3 and Figure 6.1a-c, exhibited that in the plot-A (control), the values of chlorophyll a, chlorophyll b, protochlorohyll and chlorophyll a/b ratio, observed at the 15 days interval of crop growth were recorded as 0.35±0.03, 0.39±0.05, 0.40±0.03 mg/pl and noticed a consistent increase up to 135th day stage of crop growth and amounted 1.12±0.10, 0.94±0.07, 1.11±0.11 mg/pl and 1.191 for the chlorophyll a, chlorophyll b, protochlorophyll and chlorophyll a/b ratio respectively.

A decline in all the chlorophyll pigments was identified at the maturity stage. The plants of plot-B (UV-B alone) were examined and noticed a reduction in the chlorophyll pigment as compared to control plot (A).The maximum inhibition of chlorophyll a, chlorophyll b, protochlorophyll and chlorophyll a/b ratio were recorded at the 30th day stage and 135th day stage and reduced by ca. 20 per cent, 17 per cent, 18 per cent, 5 per cent; ca. 6 per cent, 7 per cent, 9 per cent and 10 per cent respectively, as compared to control (plot-A). When the plot C, D and E were studied, with UV-B exposure along with PGRs, a general promotion in content of chlorophyll a, chlorophyll b, protochlorophyll and chlorophyll a/b ratio were recorded. Plot-C, showed the maximum amount of chlorophyll content as chlorophyll a, chlorophyll b and protochlorophyll at the 15th day stage of crop growth and a/b ratio at the maturity stage of crop growth and estimated as ca. 19 per cent, 45 per cent, 5 per cent and 20 per cent respectively, as compared to UV-B exposure alone. The plot-D was shown the maximum amount of chlorophyll a, chlorophyll b, protochlorophyll and a/b ratio observed at the 60th day and increased by ca. 6 per cent, 9 per cent, 6 per cent, 4 per cent; at 90th day ca. 7 per cent, 5 per cent, 6 per cent, 7 per cent and at maturity ca. 6 per cent, 8 per cent, 17 per cent, 11 per cent respectively, as compared to UV-B

Table 6.3: Chlorophyll contents as affected by UV-B radiation (3 hrs. daily), individually and in combination of IAA, Kn and GA_3 in field grown crop of *Brassica compestris PT-303.*

Treat-ments	*Chloro-phyll*	*15*	*30*	*45*	*60*	*75*	*90*	*105*	*120*	*135*
A	Chlorophyll a	0.35±0.03	0.47±0.03	0.058±0.04	0.065±0.07	0.097±0.08	1.04±0.04	1.06±0.06	1.09±0.08	1.12±0.10
	Chlorophyll b	0.39±0.05	0.53±0.04	0.063±0.03	0.072±0.08	0.088±0.05	0.97±0.04	0.94±0.04	0.84±0.05	0.94±0.07
	Protochlorophyll	0.40±0.03	0.50±0.02	0.068±0.04	0.68±0.08	0.83±0.05	0.086±0.07	0.91±0.01	0.99±0.01	1.11±0.11
	a/b ratio	0.89	8.87	0.92	0.902	1.102	1.072	1.127	1.297	1.191
B	Chlorophyll a	0.27±0.05	0.38±0.05	0.48±0.03	0.060±0.01	0.83±0.02	0.90±0.05	1.01±0.05	1.09±0.06	1.06±0.10
	Chlorophyll b	0.24±0.03	0.44±0.04	0.53±0.05	0.68±0.03	0.75±0.04	0.84±0.04	0.89±0.02	0.072±0.03	0.89±0.02
	Protochlorophyll	0.38±0.04	0.46±0.06	0.58±0.06	0.61±0.03	0.72±0.03	0.78±0.05	0.81±0.04	0.74±0.03	0.91±0.04
	a/b ratio	1.125	0.86	0.905	0.882	1.106	1.114	1.134	1.513	1.235
C	Chlorophyll a	0.32±0.02	0.42±0.09	0.52±0.06	0.62±0.03	0.89±0.04	0.99±0.09	1.04±0.11	1.05±0.05	1.09±0.09
	Chlorophyll b	0.35±0.03	0.46±0.06	0.60±0.05	0.70±0.07	0.85±0.08	0.91±0.01	0.90±0.08	0.89±0.09	0.99±0.07
	Protochlorophyll	0.39±0.05	0.50±0.03	0.65±0.04	0.63±0.04	0.78±0.06	0.82±0.08	0.88±0.07	1.02±0.01	1.09±0.09
	a/b ratio	0.914	0.913	0.866	0.885	1.04	1.08	1.15	1.17	1.10
D	Chlorophyll a	0.34±0.03	0.44±0.04	0.51±0.05	0.63±0.04	0.87±0.3	0.97±0.07	1.03±0.13	1.04±0.11	1.08±0.01
	Chlorophyll b	0.36±0.04	0.47±0.07	0.59±0.04	0.69±0.09	0.83±0.08	0.90±0.09	0.89±0.09	0.88±0.08	0.97±0.06
	Protochlorophyll	0.38±0.06	0.49±0.08	0.64±0.03	0.62±0.02	0.76±0.05	0.81±0.01	0.88±0.08	0.91±0.99	1.07±0.06
	a/b ratio	0.944	0.936	0.864	0.913	1.04	1.07	1.15	1.18	1.11
E	Chlorophyll a	0.33±0.03	0.44±0.04	0.50±0.05	0.62±0.09	0.88±0.08	0.98±0.09	1.02±0.12	1.05±0.22	1.09±0.02
	Chlorophyll b	0.36±0.06	0.48±0.07	0.61±0.01	0.68±0.07	0.84±0.04	0.91±0.01	0.90±0.09	0.89±0.21	0.98±0.07
	Protochlorophyll	0.37±0.07	0.49±0.08	0.63±0.03	0.61±0.06	0.75±0.05	0.82±0.02	0.87±0.07	0.92±0.03	1.06±0.06
	a/b ratio	0.91	0.92	0.81	0.91	1.04	1.07	1.13	1.17	1.11

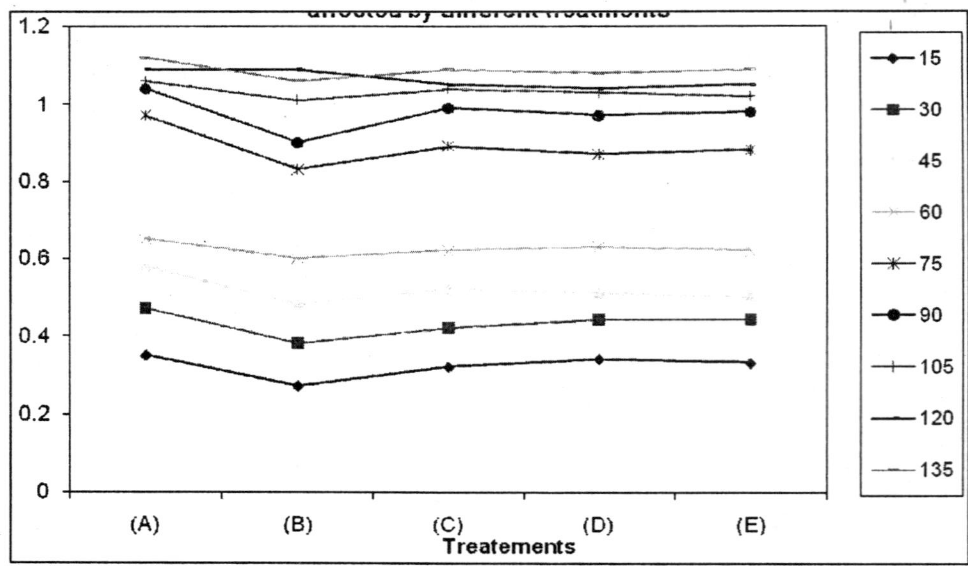

Figure 6.1(a): Chlorophyll a (mg/g) of *Brassica compestris PT-303* as affected by different treatments.

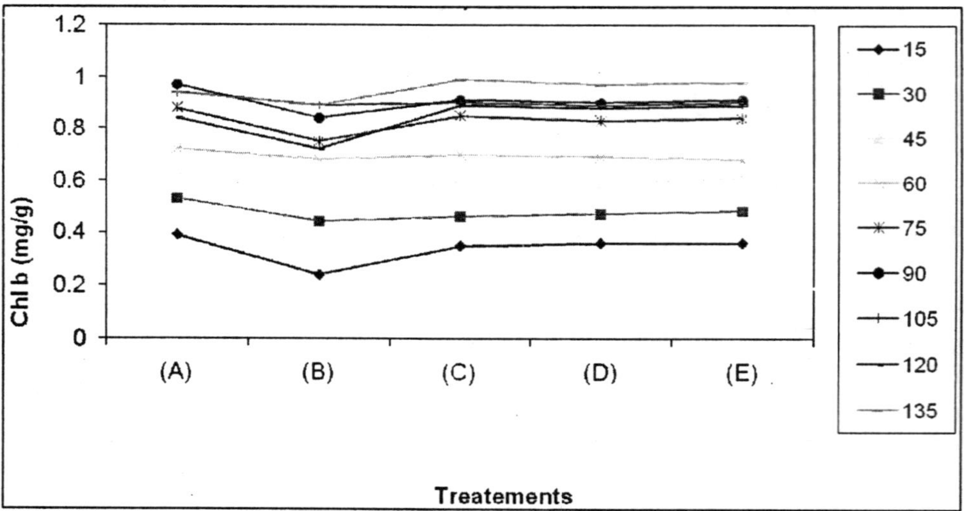

Figure 6.1(b): Chlorophyll b (mg/g) of *Brassica compestris PT-303* as affected by different treatments.

treatment. The plot-E also showed the maximum promotion of chlorophyll-a, chlorophyll-b, protochlorophyll contents and chlorophyll a/b ratio were noticed at the 30th day stage of growth and promoted by ca. 15 per cent, 10 per cent, 7 per cent, 7 per cent; at the 90th day ca.9 per cent, 5 per cent, 6 per cent, 4 per cent and at the maturity as ca. 4 per cent, 10 per cent, 16 per cent, 10 per cent respectively, as compared to the over UV-B exposure.

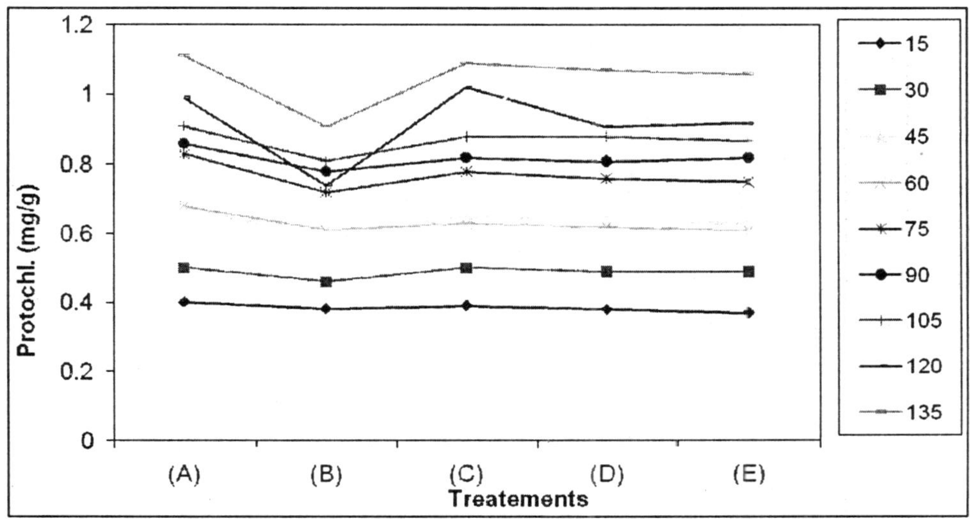

Figure 6.1(c): Protochlorophyll b (mg/g) of *Brassica compestris PT-303* as affected by different treatments.

The perusal of data presented in Table 6.4 and Figures 6.2a-c showed that 3 hrs. daily UV-B irradiation was supplied alone and along with different plant growth regulators (PGRs) affected the different chlorophyll pigments in *Brassica juncea PR-15*. The chlorophyll a, chlorophyll b, protocholorophyll and chlorophyll a/b ratio were noticed as 0.45±0.04, 0.52±0.02, 0.67±0.07 and 0.86 respectively in the control (plot A) at the 15 day interval of crop growth and increased continuously upto maturity and amounted as 1.25±0.19, 1.20±0.12, 1.21±0.08 mg/pl and 1.04 for all the chlorophyll pigments as the crop mature. The plot-B, showed a marked reduction in all chlorophyll pigments as compared to control plot. The maximum inhibition in the chlorophyll a, b, protochlorophyll and chlorophyll a/b ratio was reported at the 30th day and decreased by ca. 15 per cent, 8 per cent, 7 per cent, 8 per cent; at 60th day ca. 26 per cent, 29 per cent, 23 per cent, 6 per cent and at 90th day ca. 13 per cent, 17 per cent, 19 per cent, 3 per cent respectively as compared to control plot. When plot C, D and E were exposed to UV-B, along with PGRs, a general promotion was observed in all the chlorophyll content upto maturity as compared to UV-B treatment plot. The IAA concentration of hormone showed the maximum promotory effects at the 15th day stage of growth and continuously increased upto maturity and promoted by ca. 14 per cent, 8 per cent, 9 per cent and 9 per cent respectively in terms of chlorophyll a, b, protochlorophyll and chlorophyll a/b ratio as compared to UV-B exposure only. When Kn and GA_3 concentrations were applied, the maximum promotion was observed in terms of chlorophyll a, chlorophyll b, protochlorophyll and chlorophyll a/b ratio at 120th day and maturity stage increased by ca. 7 per cent, 6.7 per cent, 11 per cent 9 per cent and 7 per cent, 6 per cent, 9 per cent, 9 per cent respectively as compared to only UV-B.

Table 6.4: Chlorophyll contents as affected by UV-B radiation (3 hrs. daily), individually and in combination of IAA, Kn and GA_3 in field grown crop of *Brassica juncea PR-15.*

Treat-ments	*Chloro-phyll*	*15*	*30*	*45*	*60*	*75*	*90*	*105*	*120*	*135*
A	Chlorophyll a	0.45±0.04	0.57±0.12	0.64±0.01	1.06±0.12	1.09±0.14	1.12±0.11	1.16±0.15	1.19±0.17	1.25±0.19
	Chlorophyll b	0.52±0.02	0.61±0.01	0.79±0.02	1.05±0.11	1.07±0.12	1.10±0.09	1.14±0.11	1.17±0.18	1.20±0.12
	Protochlorophyll	0.67±0.07	0.73±0.03	0.89±0.09	1.09±0.15	1.08±0.11	1.11±0.08	1.15±0.12	1.16±0.09	1.21±0.08
	a/b ratio	0.86	0.93	0.81	1.01	1.02	1.08	1.09	1.1	1.04
B	Chlorophyll a	0.35±0.05	0.49±0.04	0.58±0.08	0.79±0.09	0.85±0.04	0.98±0.01	1.03±0.09	1.09±0.11	1.15±0.13
	Chlorophyll b	0.45±0.03	0.57±0.07	0.65±0.05	0.75±0.07	0.84±0.02	0.92±0.03	1.01±0.03	1.08±0.09	1.11±0.70
	Protochlorophyll	0.59±0.09	0.68±0.08	0.78±0.08	0.85±0.05	0.90±0.09	0.91±0.04	1.05±0.04	1.06±0.05	1.10±0.50
	a/b ratio	0.77	0.86	0.89	1.05	1.06	1.06	1.09	1.1	1.03
C	Chlorophyll a	0.40±0.01	0.55±0.13	0.60±0.06	0.95±0.03	0.99±0.09	1.09±0.15	1.13±0.12	1.015±0.16	1.21±0.15
	Chlorophyll b	0.49±0.05	0.59±0.01	0.69±0.07	0.89±0.09	0.94±0.04	1.08±0.13	1.11±0.09	1.16±0.12	1.16±0.11
	Protochlorophyll	0.63±0.03	0.69±0.09	0.85±0.05	0.92±0.02	0.95±0.05	1.09±0.12	1.12±0.08	1.17±0.13	1.17±0.07
	a/b ratio	0.81	0.93	0.83	1.06	1.05	1.1	1.12	0.99	1.04
D	Chlorophyll a	0.39±0.02	0.56±0.14	0.62±0.05	0.96±0.04	0.98±0.08	1.09±0.14	1.14±0.13	1.16±0.17	1.22±0.16
	Chlorophyll b	0.48±0.04	0.59±0.02	0.74±0.06	0.89±0.08	0.93±0.04	1.08±0.12	1.12±0.09	1.15±0.13	1.17±0.12
	Protochlorophyll	0.62±0.02	0.68±0.09	0.86±0.06	0.93±0.03	0.96±0.05	1.10±0.11	1.13±0.07	1.18±0.14	1.18±0.09
	a/b ratio	0.81	0.94	0.83	1.07	1.05	1.09	1.11	1.08	1.1
E	Chlorophyll a	0.40±0.04	0.58±0.15	0.63±0.06	0.97±0.07	0.99±0.09	1.08±0.15	1.15±0.01	1.17±0.18	1.23±0.17
	Chlorophyll b	0.49±0.09	0.60±0.03	0.75±0.05	0.89±0.09	0.94±0.04	1.09±0.13	1.13±0.09	1.16±0.14	1.18±0.13
	Protochlorophyll	0.63±0.03	0.67±0.08	0.87±0.07	0.94±0.04	0.95±0.03	1.11±0.11	1.14±0.04	1.19±0.13	1.19±0.07
	a/b ratio	0.81	0.96	0.84	1.08	1.06	0.99	1.01	1.08	1.04

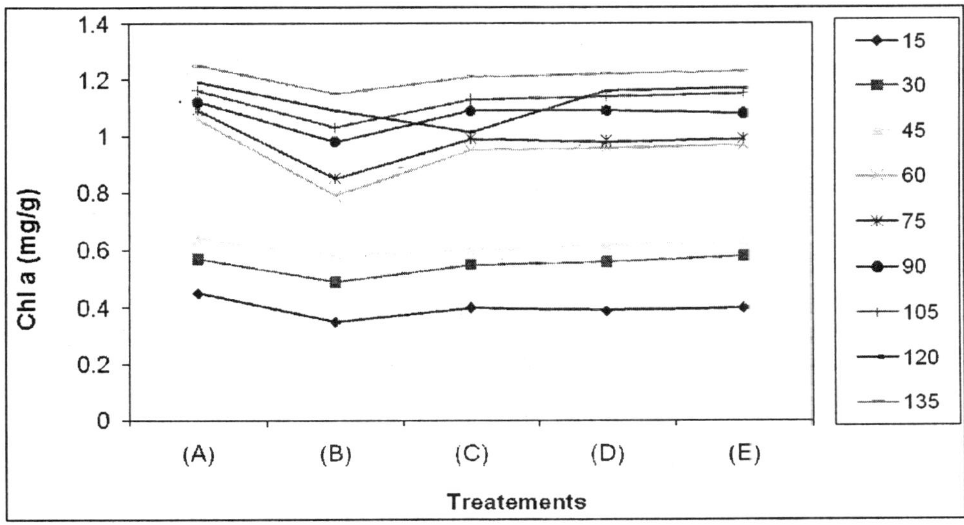

Figure 6.2(a): Chlorophyll b (mg/g) of *Brassica juncea PR-15* as affected by different treatments.

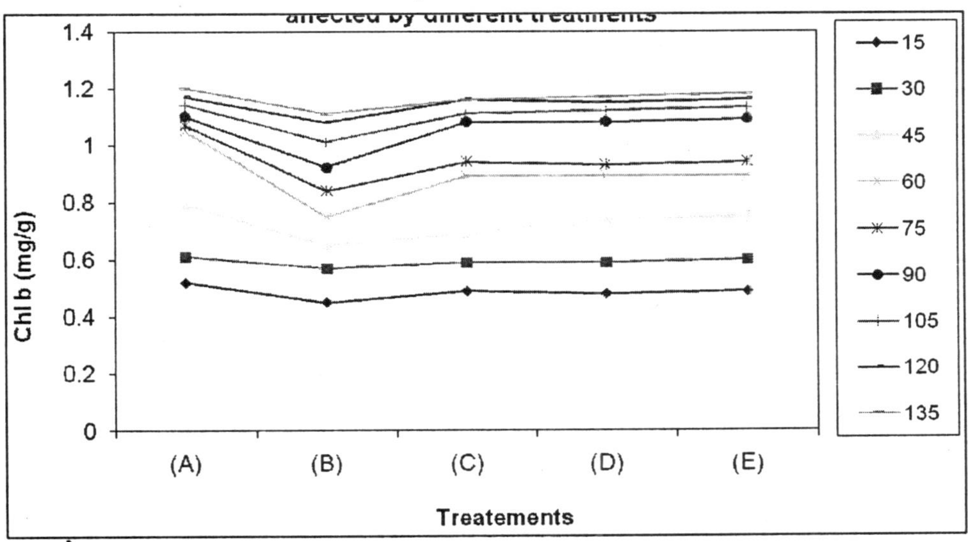

Figure 6.2(b): Chlorophyll b (mg/g) of *Brassica juncea PR-15* as affected by different treatments.

Anthocyanins

The impacts of the UV-B radiation alone and along with of IAA, Kn and GA_3 concentrations of growth regulators, on the development of anthocyanin was studied in both the varieties of mustard crops *viz.* Brown sarson and Rai. The seeds of both varieties of mustard crops were presoaked in the distilled water in dark for the 24 hrs. and then transferred in various Petridishes for the seed germination and further

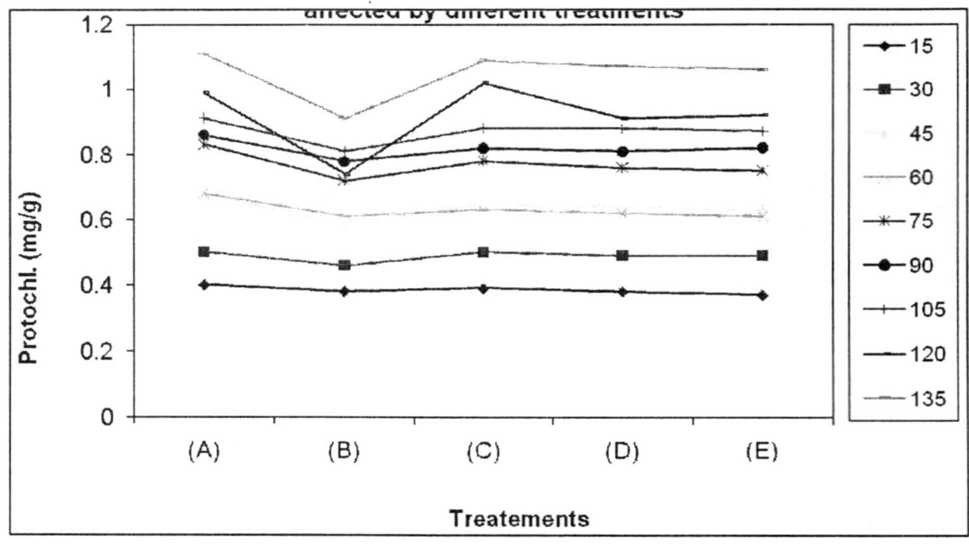

Figure 6.2(c): Protochlorophyll b (mg/g) of *Brassica juncea PR-15* as affected by different treatments.

growth. One Petridish carrying seeds of each crop was irradiated to normal white light and treated as control set. One Petridish of each variety was irradiated to 3-hrs. daily UV-B only. Three Petridishes of each crop of both varieties of mustard crops were exposed to UV-B, along with certain plant growth regulators (PGRs), carried out in the growth chamber. The seventh days old seedlings were taken for the extraction of anthocyanin as mentioned in the materials and methods.

In *Brassica compestris PT-303* and *Brassica juncea PR-15* seedlings, ultraviolet-B radiation has positive impacts on the anthocyanin accumulation. Different plant growth regulators such as IAA (10^{-7}M), Kn (10^{-5}M) and GA_3 (10^{-6}M) concentrations were applied in *Brassica compestris PT-303* and GA_3 (10^{-7}M) in *Brassica juncea PR-15* respectively and inhibited the accumulation of anthocyanin induced by UV-B exposure. A perusal of data given in Table 6.5 showed that the accumulation of anthocyanin is directly related to the UV-B exposure. In Brown Sarson 3 hrs. daily UV-B exposure caused a marked accumulation of anthocyanin as ca.128 per cent as compared to control. When IAA, Kn and GA_3 concentrations applied along with UV-B exposure were found inhibitory to anthocyanin pigment accumulation as compared to UV-B exposure alone. IAA was also found to be effective and inhibit the anthocyanin accumulation and inhibited by ca.18.7 per cent as compared to individual exposure of UV-B radiation. In case of Kn and GA_3, concentrations, it was inhibited by ca.14.2 per cent and 13 per cent as compared to UV-B exposed only.

In *Brassica juncea PR-15* (Rai), similar results were reported as in Brown sarson. Three hours daily UV-B exposure caused the marked promotion in the anthocyanin accumulation. A promotion in anythocyanin pigment was showed and recorded as ca.153 per cent as compared to the control. When different plant growth regulators such as IAA, Kn and GA_3 concentrations given along with3 hrs. daily UV-B exposure it was observed to be effective and inhibit anthocyanin accumulation level. IAA PGR was found much effective to inhibit the accumulation of this pigment and recorded as ca. 20 per cent as compared to over UV-B exposure. Kn and GA_3 concentrations also caused inhibition and recorded as ca. 26 per cent and 23 per cent of anthocyanin accumulation as compared to the individual exposure of UV-B radiation.

Table 6.5: Anthocyanin content as affected by UV-B radiation (3 hrs. daily), individually and in combination of IAA, Kn and GA_3 in field grown crops of *Brassica compestris PT-303* and *Brassica juncea PR-15.*

Treatments	*Brassica compestris PT-303*	*Brassica juncea PR-15*
A	**0.45±**0.08	**0.41±**0.07
B	**0.58±**0.06	**0.63±**0.08
C	**0.48±**0.07	**0.51±**0.06
D	**0.50±**0.05	**0.47±**0.04
E	**0.51±**0.06	**0.49±**0.09

Enzymes

Protease

The uniformly seeds of the two varieties of mustard crops were selected and soaked in the distilled water for 6 hrs, 12 hrs. and 24 hrs. respectively. After some time these presoaked seeds were spread in the different Petridishes as (A, B, C, D and E). Two Petridishes for each crop were kept as control condition, *i.e.* neither exposed to UV-B and nor PGRs, another two Petridishes were irradiated to 3 hrs. daily UV-B exposure only and nine sets of Petridishes were added with IAA, Kn and GA_3 along with UV-B radiation (3 hrs daily), for the each variety of mustard seeds. After giving the different treatments to seeds, the development of protease activity was measured and compared as described in the materials and methods.

In the seeds of the *Brassica compestris PT-303* (Table 6.6), the effect of UV-B exposure alone and along with different plant growth regulators (PGRs) on the protease activity was studied. After imbibitions of seeds in water, there was a considerable rise in the activity of protease enzyme. Data obtained from UV-B exposed Petridish, showed a marked promotion and recorded as ca. 115 per cent, 124 per cent and 120 per cent respectively at the 6 hrs, 12 hrs. and 24 hrs. as compared to control (plot-A). The

Table 6.6: Protease activity as affected by UV-B radiation (3 hrs. daily), individually and in combination of IAA, Kn and GA_3 during seed imbibitions in *Brassica compestris PT-303.*

Stage	*A*	*B*	*C*	*D*	*E*
Dry	**4.560±**0.05	--	--	--	-
6-hrs.	**8.690±**0.06	**10.070±**0.80	**13.089±**0.80	**9.080±**0.80	**11.790±**0.80
12-hrs.	**11.740±**0.07	**14.630±**0.60	**17.560±**0.230	**12.480±**0.40	**14.090±**0.49
24-hrs.	**13.870±**0.08	**16.660±**0.03	**18.650±**0.650	**13.540±**0.50	**16.430±**0.43

maximum inhibition was reported in seeds soaked in Kn and a reduction of ca.11 per cent, 15 per cent, and 19 per cent was reported at the 6 hrs, 12 hrs. and 24 hrs. respectively, as compared to the individual treatment of UV-B exposure. Protease analysis of seeds of Petridishesh IAA and GA_3 PGR_S, showed slight inhibition of protease activity as compared to the individual treatment of UV-B.

In *Brassica juncea PR-15* (Table 6.7), the effect of UV-B exposure alone and along with some plant growth regulators such as IAA, Kn and GA_3 on the protease activity. After imbibitions of seeds in water, there was a considerable rise in the activity of protease enzyme. The result was obtained from the Petridish-A (control) and recorded as ca. 8.77±0.070, 9.727±0.090, and 11.289±0.038 at the 6 hrs, 12 hrs. and 24 hrs. respectively. UV-B exposed Petridish was showed a marked promotion and increase by ca.114 per cent, 122 per cent and 139 per cent at the 6 hrs, 12 hrs. and 24 hrs. respectively, as compared to control (A). The Petridish-C showed a rise of ca. 11 per cent, 5 per cent, 6 per cent at the 6 hrs, 12 hrs. and 24 hrs. respectively as compared to individual treatment of UV-B exposure. Petridish-D showed a rise of ca. 22 per cent, 17 per cent, 2 per cent at the 6 hrs, 12 hrs. and 24 hrs. respectively as compared to individual treatment of UV-B exposure. Petridish-E also showed a rise of ca 40 per cent, 13 per cent, 31 per cent at the 6 hrs, 12 hrs. and 24 hrs. respectively as compared to individual treatment of UV-B exposure.

Table 6.7: Protease activity as affected by UV-B radiation (3 hrs. daily), individually and in combination of IAA, Kn and GA_3 during seed imbibitions in *Brassica juncea PR-15.*

Stage	*A*	*B*	*C*	*D*	*E*
Dry	0.502±0.05	–	–	--	–
6-hrs.	0.526±0.04	0.556±0.05	0.597±0.09	0.599±0.09	0.610±0.10
12-hrs.	0.590±0.09	0.630±0.03	0.780±0.08	0.820±0.07	0.850±0.05
24-hrs.	0.598±0.08	0.680±0.08	0.825±0.02	0.930±0.03	0.980±0.98

Peroxidase

The effects of UV-B exposure alone and along with of PGRs such as IAA, Kn and GA_3 on the peroxidase activity were investigated on *Brassica compestris PT-303* and *Brassica juncea PR-15* during the study of seed imbibitions.

Table 6.8 (Brown Sarson) showed that there was a considerable rise in peroxidase activity in control and noticed as ca. 0.42±0.02, 0.44±0.03 and 0.48±0.005 at the 6 hrs, 12 hrs. and 24 hrs. respectively. Petridish-B (UV-B only), showed a rise in peroxidase activity and it was reported as ca. 68 per cent, 123 per cent and 108 per cent at the 6 hrs, 12 hrs. and 24 hrs. respectively as compared to control. Petridish-C showed a rise of ca. 3 per cent, 41 per cent, 24 per cent at the 6 hrs, 12 hrs. and 24 hrs. respectively as compared to individual treatment of UV-B exposure. Petridish-D showed as ca. 2 per cent 12 per cent, 37 per cent at the 6 hrs, 12 hrs. and 24 hrs. respectively as compared to over UV-B treatment. Petridish-E was showed a rise of ca. 49 per cent, 42 per cent, 24 per cent at the 6 hrs, 12 hrs. and 24 hrs. respectively as compared to over UV-B treatment.

Table 6.8: Peroxidase activity as affected by UV-B radiation (3 hrs. daily), individually and in combination of IAA, Kn and GA_3 during seed imbibitions in *Brassica compestris PT-303.*

Stage	*A*	*B*	*C*	*D*	*E*
Dry	**0.405**±0.04	–	–	–	–
6-hrs.	**0.425**±0.02	**0.291**±0.069	**0.297**±0.07	**0.294**±0.02	**0.290**±0.03
12-hrs.	**0.445**±0.03	**0.549**±0.052	**0.775**±0.067	**0.620**±0.01	**0.780**±0.08
24-hrs.	**0.485**±0.05	**0.520**±0.030	**0.649**±0.050	**0.715**±0.052	**0.649**±0.06

Table 6.9, exhibited that peroxidsae enzyme activity was continuously increased in case of *Brassica juncea PR-15,* when subjected to individual treatment of UV-B. This peroxidase activity was reported in control as 0.525±0.04, 0.590±0.09, and 0.598±0.08 at the 6 hrs, 12 hrs. and 24 hrs. respectively. UV-B (3 hrs. daily), exposed Petridish (B), was studied and recorded an increase of ca. 105 per cent, 106 per cent and 113 per cent at the 6 hrs, 12 hrs. and 24 hrs. respectively as compared to control (A). The Petridish-C studied and recorded as increase of ca. 7 per cent, 23 per cent, 21 per cent at the 6 hrs, 12 hrs and 24 hrs respectively as compared to over UV-B treatment. Petridish-D showed an increase of ca.7 per cent, 30 per cent, 36 per cent at the 6 hrs, 12 hrs and and 24 hrs. respectively as compared to individual treatment of UV-B exposure. Petridish-E showed an increase of ca. 9 per cent, 34 per cent, 44 per cent at the 6 hrs, 12 hrs and and 24 hrs. respectively as compared to over UV-B exposure.

Table 6.9: Peroxidase activity as affected by UV-B radiation (3 hrs. daily), individually and in combination of IAA, Kn and GA_3, during seed imbibitions in *Brassica juncea PR-15.*

Stage	*A*	*B*	*C*	*D*	*E*
Dry	**0.502**±0.05	–	–	–	–
6-hrs.	**0.526**±0.04	**0.556**±0.05	**0.597**±0.09	**0.599**±0.09	**0.610**±0.01
12-hrs.	**0.590**±0.09	**0.630**±0.03	**0.780**±0.08	**0.820**±0.07	**0.850**±0.05
24-hrs.	**0.598**±0.08	**0.680**±0.08	**0.825**±0.02	**0.930**±0.03	**0.980**±0.98

Chapter 7

Discussion

The present investigation has been carried out to study the inhibitory effects of UV-B radiation and mitigatory effects of different growth regulators when applied with UV-B radiation on growth and development, biomass production, yield and some physiological aspects of *Brassica compestris PT-303* and *Brassica juncea PR-15* varieties of the mustard crops. It has been observed that when growth regulators were applied individually to germinating seeds, they enhance the growth and development of seedlings as compared to control. When UV-B was applied individually (3 hrs. daily), it reduces the growth and development of seedlings significantly. An interesting feature of the results is that when both the treatments were given to geminating seeds, pronounced mitigatory effects of growth regulators over UV-B radiation has been observed. The similar observation were also identified by Pal *et al.* (1999), Ravindran (2001), Scantos *et al.* (2004), N. Bhatt *et al.* (2004), Roa and Kushwah (2005).

Various workers in different crop plants have also reported the promotory effects of growth regulators on seed germination and survival percentage. Russell *et al.* (1998) observed the enhancing impact of Kinetin and IAA in *Phaseolus vulgaries.* They reported that Kn enhanced the germination percentage of crop significantly. Results of the present study showed 20 per cent and 24 per cent enhancement in germination when IAA was given in (10^{-7}M) concentration and 8 per cent and 5 per cent enhancement in case of Kn (10^{-5}M) concentration and 29 per cent and 23 per cent in case of GA_3 (10^{-6} M) for *Brassica compestris PT-303.* In case of *Brassica juncea PR-15,* it showed as ca. 29 per cent and 27 per cent enhancement in the germination and survival percentage when IAA was given in (10^{-7} M) concentration, 26 per cent and 29 per cent in case of Kn (10^{-5} M) concentration and 13 per cent and 6 per cent enhancement in case of GA_3 (10^{-7} M) concentration when applied individually and in combination of UV-B exposure. These results are in conformity of A. Kumar (1992), Sengupta (1999), Srivastava (1999), Jain *et al.* (1999) and N. Bhatt *et al.* (2004).

UV-B radiation (3 hrs. daily) were found to show pronounced deleterious effects on the seed germination and survival percentage of geminating seeds of both the varieties of mustard crops under investigation. The germination and survival percentage were found to decreased by 22 per cent and 16 per cent in case of *Brassica compestris PT-303* and 19 per cent and 15 per cent in *Brassica juncea PR-15* respectively when exposed to individual treatment of UV-B (3 hrs. daily). These results are with in conformity of A. Kumar (1992), while working on the effects of UV-B exposure on germination and survival percentage of *Cicer arietinum* had observed the similar results. N. Bhatt *et al.* (2004) had also reported the parallel observations on the solanaceous plant *(Lycopersicum esculentum)*. Hong-Yi Li *et al.* (2007) also observed the similar findings on the *Medicago sativa* and lettuce in which germination was significantly reduced under enhanced UV-B radiation.

When both the varieties of mustard crops were treated with UV-B radiation in combination of different concentration of IAA, Kn and of GA_3, a promotory effect of these growth regulators over UV-B radiation were noted in seedling growth. The IAA, Kn and GA_3 found to be best effective on germination and survival percentage in the concentration of (10^{-7} M), (10^{-5} M) and (10^{-6} M) in *Brassica compestris PT-303* and (10^{-7} M) in *Brassica juncea PR-15* respectively. As noted, the individual treatment of UV-B causes inhibition of germination and survival percentage of seedlings of both the varieties of mustard crops. Treatment with IAA, Kn and GA_3, showed a promotion or enhancement in same concentrations of PGR_S. The same concentration would show the maximum mitigation against UV-B induced inhibitory effects on the plants. There results are in the conformity with the observation of Pal *et al.* (1997) and Russell *et al.* (1998).

The plant growth regulators (PGR_S) play a regulatory role in the imposition of seed dormancy, in the release of seed from dormancy and in the reserve mobilization during seed germination and in the subsequent development. The promotory effects of the plant growth regulators (PGR_S), in the small concentrations are a well-known feature. The IAA has more pronounced effects on the above ground plant parts *i.e.* epicotyl and hypocotyl, whereas Kn stimulates the root growth and GA_3 also stimulates the growth of above ground plant parts. In the present investigation, when the seedlings were treated with growth regulators alone, a maximum promotory effect of ca. 45 per cent 70 per cent, 50 per cent in length, fresh weight and dry weight of epicotyl was reported. When the effect of UV-B along with IAA was studied, an increase was observed in length, fresh and dry weight of epicotyl and amounted ca. 93 per cent, 94 per cent, and 97.4 per cent; of hypocotyl ca. 69.3 per cent, 95 per cent, 94.1 per cent and of radicle ca. 90 per cent, 97 per cent, 94.4 per cent respectively as compared to UV-B exposure only. When effect of UV-B along with Kn was studied, an increase was observed in length, fresh and dry weight of epicotyl and noticed ca. 87.2 per cent, 94.2 per cent, 94.7 per cent; of hypocotyl ca. 70 per cent, 93.5 per cent, 93 per cent and of radicle ca. 87.7 per cent, 94 per cent, 80 per cent respectively as compared to UV-B exposure only. When effect of UV-B along with GA_3 was studied, an increase was also observed in length, fresh and dry weight of epicotyl and recorded ca. 87.2 per cent, 98 per cent, 92 per cent; of hypocotyl ca. 71 per cent, 95 per cent, 91 per cent and of radicle ca. 87 per cent, 98 per cent, 80 per cent respectively in *Brassica compestris PT-303* as compared to UV-B treatment only.

A similar result was also reported in *Brassica juncea PR-15,* when UV-B was given along with plant growth regulators. When effect of UV-B along with IAA was studied, an increase was observed in length, fresh and dry weight of epicotyl and recorded ca. 90 per cent, 88 per cent, 9 per cent; of hypocotyl ca. 69 per cent, 87.5 per cent, 85 per cent and of radicle ca. 76.4 per cent, 84.6 per cent, 83.7 per cent respectively as compared to UV-B treatment only. When effect of UV-B along with Kn was studied, an increase was observed in length, fresh and dry weight of epicotyl and recorded ca. 86 per cent, 88 per cent, 72.7 per cent; of hypocotyl ca. 69.8 per cent, 92.1 per cent, 78 per cent and of radicle ca. 88 per cent, 93 per cent, 74.4 per cent respectively as compared to UV-B treatment only. When effect of UV-B along with GA_3 was studied, an increase was also observed in length, fresh and dry weight of epicotyl and recorded ca. 84.7 per cent, 84 per cent, 68.1 per cent : of hypocotyl ca. 69.6 per cent, 92 per cent, 78 per cent and of radical ca. 88 per cent, 92 per cent, 90 per cent respectively as compared to UV-B exposure only. The similar results were also observed in groundnut by Kumar (1981) in *Cucumis* by Krishnamoorthy (1972) and in *Lycopersicum esculentum* by N. Bhatt *et al.* (2004).

The deleterious effects of UV-B radiation on the seedlings growth have been observed by various workers (Sullivan and Teramura, 1989, 1992) and Naidu *et al.* (1993). Alexieva *et al.* (2001) have been reported reduction in the fresh and dry biomass of wheat and pea seedlings. The soybean seedlings were also stressed, after exposure to enhanced UV-B radiation as reported by Yan, Shengrong *et al.* (2007). The high doses of UV-B radiation have been shown to induced effects in coniferous, including reductions in the seedling biomass and height. The germination percentage was noted to decrease by 15 per cent in *Vigna mungo* and 8 per cent in *Vigna aureus* crops respectively when crops were treated with UV-B radiation as observed by Ambrish (1992). The effects of UV-B and relative nutrient addition rate were mostly additive rather then synergistic effects on growth of *Betula pendula* seedlings as observed by Rosa *et al.* (2001). The effects of the UV-B radiation were observed on the growth and development of cucumber seedlings. The UV-B radiation induces range of effects in plants which include leaf thickening, cotyledon curling and inhibition of hypocotyl, stem and leaf elongation, axillary branching and shifts in the root, shoot ratio were observed by Barnes *et al.* (1990), Cen and Bornman (1993), Ballare *et al.* (1995), Jansen *et al.* (1998), Kim *et al.* (1998) and Boccalandro *et al.* (2001).

In the present investigation, a reduction of ca. 75 per cent, 62 per cent and 76 per cent was reported in epicotyl, hypocotyl and radicle length of *Brassica compestris PT-303.* In case of *Brassica juncea PR-15, the* length of epicotyl, hypocotyl and radicle were reduced by ca. 73 per cent, 62.2 per cent and 76 per cent respectively. Similar results of decrease in fresh and dry weight were reported for both the varieties of mustard crops. N. Bhatt *et al.* (2004) also observed the similar results while working on the tomato plant. Under present investigation, all parts of the plants were observed susceptible to the enhanced level of UV-B radiation.

Generally, combined effect, of UV-B and different growth regulators were found promotory when compared with the UV-B treated seedlings of both the varieties of mustard crops. Epicotyl, hypocotyl and radicle length, fresh and dry weight has been found increased over UV-B individual treatment. IAA (10^{-7}M) and Kn (10^{-5} M) and

GA_3 (10^{-6} M) in *Brassica compestris PT-303* and (10^{-7}M) in *Brassica juncea PR-15*, were found most effective in case of epicotyl, hypocotyl and radicle respectively. While the growth regulators have their pronounced promotory effects on different parts of the seedling as reported here and elsewhere and UV-B suppress the development of different parts of the seedlings. The combined treatment of UV-B and PGRs showed their promotory effect as compared to individual treatment of UV-B. Similar results have also been observed by A. Kumar (1992) and Russell *et al.* (1998).

In general individual treatment of UV-B radiation showed pronounced inhibitory effects on growth in terms of leaf area, fresh weight and dry weight of leaves, root, stem, flower, fruit and their fresh and dry weight in both the varieties of mustard crops. When crops under investigation were treated with UV-B in combination of PGRs, an increase in all the above parameters were observed as compared to individual treatment of UV-B radiation.

A number of studies have been carried out in India and abroad on the effects of UV-B radiation on growth and development of various crops and natural vegetation as observed by Murali &Teramura (1985); Goyal and Jain (1990); Kulandaivelu, G. and K. Annamalainathan (1991); Ambrish (1992); Baker *et al.*,(1994); Ambashat *et al.* (1995); Barsing *et al.* (2000); Smith *et al.* (2000); Alexieva *et al.* (2001); Levizov and Manetas, (2001); Agrawal *et al.* (2004) and N. Bhatt (2004). Daily treatment of UV-B exposure (3 hrs. daily) observed the overall inhibition in number and weight of all parts of plants. The reduction was observed in plant height and dry weight of soyabean, when subjected to UV-B radiation. The plant height and total dry weight of soyabean was decreased by ca. 56 per cent and 42 per cent respectively caused by UV-B radiation as reported by Biggs *et al.* (1981).

Fruiting determine the yield of a crop. Hart *et al.* (1975) observed a significant reduction in fruit in pepper as influenced by UV-B radiation. Biggs and Kossuth (1978) and Ambrish, (1992) were reported similar results for flower and fruit development. Ohtani *et al.* (1982) reported inhibition of flowering due to blue light near UV-B radiation in *Lemna paucicostata*. Flowering was also found inhibited in rice during its kernel development due to UV-B radiation by Morie *et al.* (1989).

When the crops were treated with UV-B in combination of different concentrations of IAA, Kn and GA_3, a promotion in all the growth parameters was observed as compared to individual treatment of UV-B in both the varieties of mustard crops. The IAA was found to be most effective in case of leaf, which enhanced ca. 10 per cent of leaf area; ca. 30 per cent of fresh weight and ca. 11 per cent of dry weight as compared to individual treatment of UV-B in *Brassica compestris PT-303* as compared to UV-B exposure only. Kn was found to increase leaf area, fresh and dry weight and enhanced ca. 7.2 per cent, 24.6 per cent and 10 per cent respectively. GA_3 was also reported to promote leaf area, fresh and dry weight and enhanced ca. 10 per cent, 27.5 per cent and 23.2 per cent respectively as compared to UV-B exposure only.

In case of *Brassica juncea PR-15*, IAA was found most effective in case of leaf area, fresh weight and dry weight and enhanced ca. 16 per cent, 16.5 per cent and 16 per cent respectively as compared to individual treatment of UV-B radiation. Kn was found most effective in terms of leaf area, fresh weight and dry weight and recorded ca. 16.6 per cent, 16.1 per cent and 34 per cent respectively as compared to individual treatment of UV-B radiation. The GA_3 was also observed most effective in terms of the leaf area, fresh weight and dry weight and recorded ca. 10.3 per cent, 3.6 per cent and 35 per cent respectively as compared to individual treatment of UV-B exposure. The above studies found support from the work of Jauhri *et al.* (1960), Biggs and Kossuth (1978), Dickson and Caldwall (1978), Kumar (1981), Ambrish (1992), Agarwal *et al.* (1991), Kaur *et al.* (1998) and Russell *et al.* (1998).

Standing crop of biomass represents long-term integration of biochemical, physiological and growth of the plants. Therefore, UV-B and PGRs individually and in combination induced the physiological processes and expresses their effects significantly in biomass partitioning. Total plant biomass was found increased consistently and linearly in all the treatments in both the varieties of mustard crops. Maximum inhibition of biomass due to continuous UV-B exposure was observed ca. 45 per cent and 39 per cent at the 105th day stage in *Brassica compestris PT-303* and *Brassica juncea PR-15* respectively. When the crops were treated with combined treatment of UV-B and PGRs, maximum mitigatory effects was reported for IAA and amounting to ca. 26.8 per cent and 32.4 per cent approximately in both varieties of mustard crops as compared to individual treatment of UV-B exposure. In case of GA_3 hormone, the maximum mitigatory effects were also reported in total standing crop of biomass and amounting to ca. 30.4 per cent and 30.8 per cent approximately for both the varieties of mustard crops as compared to individual treatment of UV-B exposure.

Various workers have shown that PGRs when applied to the crop individually enhance the accumulation of dry matter biomass and at the same time, UV-B reduces it substantially. Biggs and Kossuth (1978) reported that total dry weight reductions in field-grown corn, pea, tomato and mustard caused by UV-B radiation. Sullivan *et al.* (1996), Cuarda *et al.* (2004) and Yang *et al.* (2005) have been observed that total biomass and leaf area decreased in response to enhanced UV-B radiation. The morphological responses induced by enhanced UV-B exposure have been attributed to hormonal imbalance resulting from the interaction between UV-B radiation and IAA metabolism (Ros and Tevini, 1995). Tevini *et al.* (2000) observed that enhanced UV-B radiation can deleteriously affect overall growth and biomass accumulation in the plant species. UV-B exposure reduced stem height due to shorter internode rather than reduced number of nodes as observed in the *Pisum sativum* (Gonzalez *et al.*, 1998). Reduction in the total biomass is a consequence of decreased plant height. The effects of UV-B exposure on plants include reduced biomass accumulation, altered biomass allocation and increased falvonoid content reported by Hofmann *et al.* (2001) and Kolb *et al.* (2001).

Possible reductions in leaf biomass, leaf area and carboxylation efficiency in response to elevated UV-B radiation would be expected to lead the reduced productivity (Keiller and Holes, 2001). The leaf growth patterns are very sensitive to environmental stress (Dillenburg *et al.*, 1995). Barisng and Malz, (2000) observed that drastic change in the starch and sucrose partitioning in the *Zea mays* due to leaf exposure to UV-B. The higher reduction in the biomass of leaves than that of stem or spikes in *Triticum aestivum* due to enhanced UV-B radiation as observed by Yue *et al.* (1998).

Fiscus and Booker (1995) reported that the direct consequences of such effects caused a decrease in crop yields of 20-25 per cent. Deleterious effects of UV-B may largely be partitioned between damages to the plant genome and to the photosynthetic machinery. Direct damage to DNA is a common result of absorption of high-energy UV-B photons. It has also been reported that in some plants under the proper conditions, almost every facet of photosynthetic machinery can be damaged directly by very high UV-B exposures. However, electron transport mediated by PS II appears to be the most sensitive part of the system. Vu *et al.* (1982) reported damage to virtually all parts of the PS II from the Mn binding site to the PQ acceptor site on the opposite surface of the thylakoid membrane.

Teramura and Sullivan (1994) observed that leaf optical properties of terrestrial plant apparently minimize the exposure of sensitive targets to UV-B radiation. Secondary effects of this damage may include reductions in photosynthetic capacity, RuBP regeneration and quantum yield. Furthermore, UV-B radiation may decrease the penetration of PAR, reduce photosynthetic and accessory pigments impair stomatal functions, alter canopy morphology, thus, resulting into reduced accumulation of dry matter biomass.

TNP of both the crops were found altered due to individual and combined effects of UV-B radiation and PGRs on the *Brassica compestris PT-303* and *Brassica juncea PR-15* respectively. Goyal and Jain (1990) observed that 3 hrs. exposure of UV-B radiation to linseed crop exhibit significant reduction in primary production of different plant parts. The similar result was found that the effects of UV-B radiation on the plant have been carried out by some other workers. Most of these studies have been carried out only to evaluate the deleterious effects of UV-B radiation on net primary productively. Kumar *et al.* (1988) reported that TNP was reduced under supplemental UV-B radiation in the field grown lentil crop. Biggs and Kossuth (1978) studied the impact of solar UV-B radiation on crop productivity. The effects of UV-B radiation on the primary production of natural phytoplankton assemblages in Michigan Lake were also reported by Gala and Giesy (1991).

When UV-B was applied with daily spray of IAA, Kn and GA_3 an enhancement in NPP of different plant part was noted in all the treatment of both the varieties of mustard crops. The increase in dry matter production may be the result of more uptakes of nutrients and synthesis of reserved food material as affected by growth regulators (Irulappan and Muthukrishnan, 1973 and Kumar, 1981).

Reports are available that application of PGRs ultimately affects the endogenous level of auxins (Andreae and Andreae, 1953; Kuraishi and Muir, 1963 and Wort, 1964), which finally affects the growth, and development of plant. Auxins interact with one or more components of the biochemical system involved in the protein synthesis. However, it has not been identified the proper step where auxins exert an effect. According to popular concept, auxins do act through influence upon enzyme production. There are definite evidences that nucleic acids are involved in growth. Roychoudhary and Sen (1964) found that application of auxins to peas resulted in an enhanced RNA synthesis.

Similarly, Key and Shanon (1964) found that the incorporation of labelled nucleotides into the nucleic acid is stimulated by auxins. Collectively, these experiments imply that the regulation of growth by auxins may involve the regulation of RNA synthesis and hence, the protein synthesis. Noggle and Fritz (1976) stated that auxins might cause the movement of more sugars into the vacuoles so that more water may enter the cell till the development of sufficient wall pressure. The above findings support the mitigatory effects or Auxins and Kn towards the deleterious effects caused by UV-B to NPP.

Harvesting index represents the economic yield or distribution of dry matter in storage parts of the plant. In the present investigation, it was reported that different treatments affect the economic yield of both the crops significantly. The UV-B radiation caused a decrease of ca. 4 per cent and 4.3 per cent in case of *Brassica compestris PT-303* and *Brassica juncea PR-15* respectively. A promotory effect was noted when UV-B was given along with Kn (10^{-5} M) and GA_3 (10^{-6} M) (3 per cent and 2.5 per cent respectively). Other growth regulators did not show any promising results on this aspect. Ambrish (1992) while working with *Cicer arietinum* reported the similar results when treated with UV-B only. Thus longer exposure of UV-B reduces the photosynthetic activity of plants and hence, the depositions of food materials in storage organs may affects as observed by Teramura, (1980) and Kumar, (1988). The reduction in the spike and grain weights reflects alteration of photosynthetic allocation due to UV-B radiation. The adverse effects could be due to reduction of photosynthesis in flag leaves of exposed plants responsible for photosynthetic allocation to economically important parts. Demchik and Day (1996) reported that the reduction in seed weight of *Brassica rapa* under 16 per cent and 32 per cent ozone depleted UV-B influx. Similarly, reduction in quantity of pollen grains *vis-a-vis* their viability under enhanced UV-B radiation might also contribute to reduction in grain yield. So this penetration of UV-B might be more effective in reducing grain set /yield.

Shelling Percentage of both the crops was affected adversely by UV-B radiation in the present investigation. When *Brassica compestris PT-303*, subjected to combined treatment of UV-B along with IAA, an increase in Shelling Percentage was reported as ca. 5 per cent. In case *of Brassica juneca PR-15,* subjected to combined treatment of UV-B along with Kn, an increase in Shelling Percentage was reported as ca. 6 per cent. The other growth regulators did not show the any promising result. Alteration in number of fruits and seed quality and ultimately the SP was also found adversely affected as observed by Ambler *et al.* (1978) and Biggs and Kossuth (1978), thus, supporting the results obtained in present study. N. Bhatt (2004) also observed that UV-B exposure significantly affected the growth and yield of certain crop plants. Since plant growth regulators (PGRs) were observed to increase the Shelling Percentage significantly, as studied by Chauhan *et al.* (1970); Badawi (1978); Ram *et al.* (1973); Shamshery and Gangwar (1979); Patil *et al.* (1987) and Kumar *et al.* (1996). Mitigation of UV-B induced inhibition on Shelling Percentage by PGRs, found in the present study is supported by the above finding by A. Kumar (1992) also reported the similar observation, while working on the certain legume crops respectively.

In the present study, carried out in the laboratory, destruction of chlorophyll, a, b, protochlorophyll and chl a/b ratio was noticed when both the crops were treated with UV-B radiation. In this crop, chlorophyll a and chlorophyll b were found almost equally reduced due to 3 hrs. daily treatment of UV-B. When the crops were supplemented with PGRs in addition to the UV-B radiation, a promotory effect was noted in the present study. The IAA and GA_3 were found most promising growth regulators, when compared with Kn. The significant reductions in different chlorophyll pigment by UV-B exposure were also investigated by Jain and Goyal (1990), Sharma *et al.* (1988), Goyal *et al.* (1991) and Ambrish (1992).

The chlorophyll content were also analyzed in the field grown crops under the influence of various treatments. In general, it was observed that UV-B inhibits the chlorophyll development throughout the crop age. However, more reduction was recorded in early stages of growth and at maturity. Kn, when applied with UV-B radiation, was found to enhance the different chlorophyll pigments level in both the varieties of mustard crops, however, the other PGRs also mitigate the adverse effects of UV-B, marginally.

These findings showed the lethal effects of UV-B towards chlorophyll development and repaired by Kn (10^{-5} M). This effect was found variable with the crop species. Vu *et al.* (1981) reported that chlorophyll a/b ratio decreased due to UV-B radiation in soybean but increased in pea. Tevini *et al.* (1981) concluded that UV-B radiation inhibited the biosynthesis of chlorophyll b than chlorophyll a. Jain and Goyal (1990), while working with lentil crop under field conditions, reported the similar results. They also emphasized that interconversion of protochlorophyll to chlorophyll was retarded. As Kn was found to improve the synthesis of chlorophyll even under increased radiation energy (Purohit, 1988), an improvement in different chlorophyll contents was reported in the present study under similar conditions.

One of the measures, which plants develop for the defence towards higher UV-B radiation, is the development of anthocyanin. Present study showed that both the crops *viz. Brassica compestris PT-303* and *Brassica juncea PR-15* develop over 128 per cent and 153 per cent anthocyanin production as compared to control when treated with 3 hrs daily UV-B radiation. A slight decrease in anthocyanin content was noted when the crops were exposed to combined effects of UV-B and different growth regulators. This shows that growth regulators caused insignificant change in the anthocyanin accumulation in plants towards UV-B radiation. Ambler *et al.* (1978) and Bennett (1981) found the accumulation of anthocyanin as a defence of cotton plants against enhanced UV-B radiation. Hashimoto *et al.* (1991) also reported the similar observation while working with chlorophyll due to enhanced UV-B radiation can be correlated with each other. Enhancement of anthocyanin synthesis can be explained as chloroplast may provide a large reserve pool for the biosynthesis of anthocyanin (Mancinelli *et al.*, 1975). UV-B induced anthocyanins production has also been reported in mustard hypocotyl, corn, wheat and rye coleoptiles (Wellmann, 1982). Arakawa *et al.* (1985) found synergistic increase in anthocyanin production caused by UV-B (312 nm) with white light in apple fruits. Yatsuhashi and Hashimoto (1985) found multi facet action of UV-B photoreceptor and phytochrome in the synthesis of anthocyanin using 290 nm (UV-B radiation).

Similar to anthocyanin, flavonoid concentration was also increased in UV-B treated seedlings after four days of treatment. In contrast, high UV-B influence increased the flavonoid accumulation (Prem Kumar *et al.*, 2001). According to (Tevini *et al.*, 1990) flavonoid accumulation is regarded as protective mechanism in higher plants to provide against UV-B radiation. Hence, it is concluded that the UV-B treated seedlings may activate a defence mechanism against UV-B damage by increasing flavonoid. Pal *et al.* (1999) concluded that flavonoid concentration can reduce the UV-B penetration and protect the photosynthetic apparatus upto some extent, but it depends upon threshold level which may vary in different species. However, there is also evidence that flavonoids may function in plants to screen harmful radiation, bind phytotoxins and help to regulate the stress response by controlling auxin transport (Shirley, 2002).

This study showed that the considerable rise in protease and peroxidase activities in the germinating seeds as compared to pre-existing enzymes in the seeds. The experimental data showed that the enhancement of protease activity upto ca. 115 per cent at 6 hrs; 124 per cent at 12 hrs. and 120 per cent at the 24 hrs. respectively in the UV-B exposed geminating seeds of *Brassica compestris PT-303* as compared to control condition. When the crop was subjected to combined treatment of UV-B along with Kn (10^{-5}M), the most mitigatory effects was found and which lowered the activity of protease upto 15 per cent at the 6 hrs; 11 per cent at the 12 hrs and 19 per cent at the 24 hrs. respectively as compared to individual treatment of UV-B exposure. Other growth regulators have showed the slight inhibition as compared to UV-B exposure alone. When the crop was exposed to UV-B only, showed a considerable rise in the peroxidase activity and reported as ca. 68 per cent, 123 per cent and 108 per cent at 6 hrs, 12 hrs. and 24 hrs. respectively as compared to control. When the crop was subjected to UV-B along with PGRs such as IAA, showed a rise of ca. 3 per cent at 6 hrs, 41 per cent at 12 hrs and 24 per cent at 24 hrs; Kn of ca. 2 per cent at 6 hrs; 12 per cent at 12 hrs; 37 per cent at 24 hrs. and GA_3 of ca. 99 per cent at 6 hrs; 42 per cent at 12 hrs; 24 per cent at 24 hrs. respectively as compared to individual treatment of UV-B exposure.

In case of *Brassica juncea PR-15*, exposed to UV-B radiation, a marked promotion was showed and increased by ca 114 per cent, 122 per cent and 139 per cent at 6 hrs, 12 hrs and 24 hrs. respectively in terms of protease activity as compared to control. When the crop was subjected to combined treatment of UV-B along with PGRs such as IAA, showed a rise of ca. 11 per cent, 5 per cent, 6 per cent; Kn of ca. 22 per cent 17 per cent, 2 per cent and GA_3 of ca. 40 per cent, 13 per cent, 31 per cent at 6 hrs, 12 hrs and 24 hrs. respectively as compared to individual treatment of UV-B exposure. When crop exposed to individual treatment of UV-B exposure, the peroxidase activity was studied and recorded an increase of ca. 105 per cent, 106 and 113 per cent at 6 hrs, 12 hrs and 24 hrs. respectively as compared to control. When crop subjected to combined treatment of UV-B along with PGRs *viz.* IAA, showed a rise of ca. 7 per cent at 6 hrs, 23 per cent at 12 hrs, 21 per cent at 24 hrs; Kn of ca. 7 per cent at 6 hrs, 30 per cent at 12 hrs, 36 per cent at 24 hrs and GA_3 of ca. 9 per cent at 6 hrs, 34 per cent at 12 hrs, 44 per cent at 24 hrs. respectively as compared to individual treatment of UV-B exposure.

Chapter 8

Summary

Seed germination, shelling percentage, growth behavior, biomass and productivity, compartmental partitioning of biomass and some physiological and biological aspects of *Brassica compestris PT-303* and *Brassica juncea PR-15* were investigated in the present research work. Under the influence of enhanced level of UV-B irradiation, individually and in the combination of some plant growth regulators (PGRs), in *Brassica compestris PT-303* and in *Brassica juncea PR-15* respectively to evaluate the mitigatory effects of these hormones against UV-B radiations. The findings were observed through this study are summarized as follows.

1. During laboratory study to evaluate the appropriate concentrations of different plant growth regulators (PGR_S) for the field study, the seeds of both varieties of mustard crops were grown in Petridishes. The seeds of *Brassica compestris PT-303* and *Brassica juncea PR-15* were treated with IAA, Kn and GA_3, in their appropriate concentrations *i.e.*(10^{-7}) to (10^{-5} M) individually and in combination of UV-B, it was observed that IAA was found most effective in (10^{-7} M), Kn in (10^{-5} M) and GA_3 in (10^{-6} M) in *Brassica compestris PT-303* and (10^{-7} M) in *Brassica juncea PR-15* respectively. Therefore, for the field studies only these concentrations were taken to assess the comparative impacts of these PGRs to mitigate UV-B induced deleterious effects.
2. Plot-A was taken as control (untreated). Plot-B was treated with UV-B radiation (3 hrs. daily). Plot-C was sprayed daily with IAA (10^{-7} M) with 3 hrs. daily UV-B exposure. Plot-D was sprayed with Kn (10^{-5} M) along with 3 hrs. daily UV-B exposure and Plot-E was also sprayed with GA_3 (10^{-6} M)) in *Brassica compestris PT-303* and (10^{-7} M) in *Brassica juncea PR-15* respectively along with 3 hrs. daily UV-B exposure. The above treatments were given to both varieties of mustard crops and plots were named simultaneously as above.

3. The germination and survival percentage for *Brassica compestris PT-303* was recorded as ca. 80 per cent and 69 per cent respectively in the control condition. This percentage was enhanced to ca. 81 per cent and 72 per cent in case of IAA (10^{-7} M); 83 per cent and 80 per cent in case of Kn (10^{-5}M) and 80 per cent and 76 per cent in case of GA_3 (10^{-6} M) respectively. This percentage was reduced to 65 per cent and 59 per cent when exposed to UV-B radiation (3 hrs. daily). Mitigatory effects of certain PGRs were observed on the crops when the seeds were treated with plant growth regulators along with UV-B radiation as IAA + UV-B recorded as ca. 78 per cent, 66 per cent; Kn + UV-B recorded as ca. 60 per cent, 56 per cent and GA_3 + UV-B recorded as ca. 84 per cent, 71 per cent for germination and survival percentage of *Brassica compestris PT-303* respectively.
4. In case of *Brassica juncea PR-15*, the similar results were also observed. The germination and survival percentage in controlled condition was amounted ca. 80 per cent and 69 per cent respectively. This percentage was enhanced and amounted ca.79 per cent and 76 per cent with IAA (10^{-7}M), 78 per cent and 75 per cent with Kn (10^{-5}M) and 79 per cent and 74 per cent with GA_3 (10^{-7} M) respectively. When UV-B exposure was given 3 hrs. daily), the germination and survival percentage was reduced and inhibited to ca. 61 per cent and 55 per cent respectively. The combined effects of PGRs, when given along with UV-B radiation daily, which enhanced germination and survival percentage as compared to individual treatment of UV-B radiation and amounted to 69 per cent, 63 per cent with IAA + UV-B; 65 per cent, 61 per cent with Kn + UV-B and 65 per cent, 61 per cent with GA_3 + UV-B respectively.
5. Stem growth patterns were also found inhibited due to both the treatments. UV-B (3 hrs. daily) reduced the stem growth considerably in terms of stem length, fresh weight and dry weight. The length, fresh and dry weight was found to be inhibited by ca. 17 per cent, 25 per cent, 31 per cent at the 15th day; ca. 19 per cent, 21 per cent, 40 per cent at the 30th day; ca.7 per cent, 26 per cent, 42 per cent at the 105th day and ca. 9 per cent, 30 per cent, 40 per cent at the maturity respectively as compared to control condition. When the crops were treated along with plant growth regulators (PGR_S) *viz.* IAA (10^{-7}M) in combination of UV-B exposure, the promotory effect was observed in terms of length, fresh weight and dry weight was noticed at the 60th day as ca. 13.4 per cent, 5.3 per cent, 4.1 per cent respectively as compared to control. When the crop was sprayed along with Kn (10^{-5} M) in combination of UV-B exposure, the maximum promotory effect was observed in terms of length, fresh weight and dry weight were noticed as ca. 3.3 per cent 32.4 per cent, 24 per cent respectively as compared to UV-B exposure. When the crop was also sprayed along with GA_3 (10^{-6} M) in combination of UV-B exposure, the maximum promotory effect was observed in terms of length, fresh weight and dry weight were noticed as ca. 5.8 per cent, 19 per cent, 62 per cent respectively as compared to UV-B exposure.

In *Brassica juncea PR-15,* the inhibition due to UV-B radiation was recorded as ca. 19 per cent, 30 per cent and 40 per cent respectively as compared to control. When the crop was exposed along with plant growth regulators in combination of UV-B radiation, the promotory effects were reported in terms of length, fresh weight and dry weight of stem observed as ca. 19 per cent, 18 per cent, 13 per cent in IAA (10^{-7} M) + UV-B; ca. 19 per cent, 21 per cent, 15 per cent in Kn (10^{-5} M) + UV-B; ca. 20 per cent 11 per cent, 24 per cent in GA_3 (10^{-7} M) + UV-B as compared to individual treatment of UV-B exposure.

6. Leaf growth patterns were found significantly inhibited due to individual treatment of UV-B exposure in terms of leaf area, fresh weight and dry weight was reduced by ca. 26 per cent, 19 per cent, 39 per cent respectively as compared to control. IAA (10^{-7} M) and GA_3 (10^{-6} M) were found maximum mitigatory effects in terms of the leaf growth patterns against UV-B damage and mitigated these parameters by ca. 10 per cent, 30 per cent, 11 per cent and 10 per cent, 27.5 per cent, 23.2 per cent respectively in *Brassica compestris PT-303.* In *Brassica juncea PR-15,* the inhibitory effect due to UV-B exposure was recorded ca. 72 per cent, 10 per cent and 21 per cent respectively in leaf growth pattern. When these concentrations of PGRs *viz.* IAA, Kn and GA_3 were applied and observed promotory effects upto 16 per cent, 16.5 per cent, 16.1 per cent; 16.6 per cent, 16.1 per cent, 34.3 per cent and 10.3 per cent, 3.6 per cent, 35 per cent respectively as compared to UV-B exposure in terms of leaf area, fresh and dry weight against UV-B exposure.
7. Similarly the root growth pattern was also found affected with both the treatments. In case of *Brassica compestris PT-303,* a reduction was amounted to ca. 20.2 per cent, 53 per cent respectively in fresh weight and dry weight of root when exposed to UV-B only. IAA was reported maximum effective to mitigate the UV-B induced root growth inhibition and increased by ca. 19 per cent and 22 per cent respectively in fresh weight and dry weight. Kn was also reported maximum effective to mitigate the UV-B induced root growth inhibition and increased by ca. 19 per cent and 31 per cent respectively in fresh weight and dry weight. In case of *Brassica juncea PR-15,* the UV-B induced inhibition was recorded ca. 52 per cent and 62 per cent in fresh weight and dry weight respectively. IAA improved the inhibition in combined treatment and promoted ca. 26.9 per cent and ca. 23 per cent over UV-B individual treatment for fresh weight and dry weight. Kn improved the inhibition in combined treatment and promoted ca. 30.6 per cent and 20 per cent as compared to UV-B individual treatment for the fresh weight and dry weight of root.
8. Flower growth pattern was also observed and affected by the both treatments. A reduction was observed and noticed as ca. 45 per cent, 65 per cent respectively in terms of fresh weight and dry weight of flowers, when exposed to individual treatment of UV-B exposure as compared to control. IAA was reported maximum effective to mitigate the UV-B induced flower growth inhibition and increased by ca. 53 per cent and 80 per cent respectively in fresh weight and dry weight of flowers as compared to

individual treatment of UV-B exposure. Kn was reported maximum effective to mitigate the UV-B induced flower growth inhibition and increased by ca. 53 per cent and 85 per cent respectively in fresh weight and dry weight of flowers as compared to individual treatment of UV-B exposure. GA_3 was also reported maximum effective to mitigate the UV-B induced flower growth inhibition and increased by ca. 13 per cent and 23 per cent respectively in fresh weight and dry weight of flowers as compared to individual treatment of UV-B exposure.

In *Brassica juncea PR-15,* the similar results were also observed. The UV-B inhibited flower growth and recorded by ca. 45 per cent and 68 per cent respectively as compared to control. In *Brassica juncea PR-15,* the similar results were also observed. IAA was reported maximum effective to mitigate the UV-B induced flower growth inhibition and increased by ca. 53 per cent and 90 per cent respectively in fresh weight and dry weight of flowers as compared to individual treatment of UV-B exposure. Kn was reported maximum effective to mitigate the UV-B induced flower growth inhibition and increased by ca. 17 per cent and 5.9 per cent respectively in fresh weight and dry weight of flowers as compared to individual treatment of UV-B exposure. GA_3 was also reported maximum effective to mitigate the UV-B induced flower growth inhibition and increased by ca. 30 per cent and 93 per cent respectively in fresh weight and dry weight of flower as compared to individual treatment of UV-B exposure.

9. Fruiting is the index of the yield of crops. It was also found affected significantly due to individual and combined treatments of UV-B with different PGRs. In case of *Brassica compestris PT-303,* the maximum inhibitions in the fresh and dry weight of fruit were amounted ca. 50 per cent and 58 per cent due to UV-B radiation. It was found improved in combination of UV-B treatment along with IAA and found to be increased by 44 per cent and 82 per cent for fresh weight and dry weight of fruit respectively as compared to individual UV-B treatment. When UV-B supplied along with Kn, it was increased by 24 per cent and 13 per cent for fresh weight and dry weight of fruit as compared to control. When UV-B supplied along with GA_3, it was also increased by 19 per cent and 20 per cent respectively for fresh weight and dry weight of fruit as compared to control. In *Brassica juncea PR-15,* the inhibition was recorded as ca. 42 per cent and 85.7 per cent respectively in fresh and dry weight of fruit as compared to control. IAA improved the deleterious effects when given in combination and improved by ca. 38.9 per cent and 34.1 per cent in fresh and dry weight of fruits respectively. Kn when given along with UV-B as compared to individual UV-B exposure brought about an improvement of ca. 42.2 per cent and 12.8 per cent in fresh and dry weight of fruit. GA_3 when given along with UV-B as compared to individual UV-B exposure brought about an improvement of ca. 26 per cent and 21 per cent in fresh and dry weight of fruit.

10. Standing crop of biomass was found increased consistently and linearly in both the crops upto maturity except leaf biomass, which was found, reduced at maturity due to leaf fall. UV-B was found affecting biomass of each component when supplied individually in both the varieties of mustard crops. The total biomass was found reduced by ca. 45 per cent as compared to control. In case of *Brassica compestris PT-303,* IAA along with UV-B radiation was showed enhancement by 26.8 per cent, along with Kn an enhancement of 27.3 per cent and along with GA_3 an enhancement of 34.4 per cent was noted as compared to UV-B alone. *Brassica juncea PR-15,* when exposed to UV-B only, the total biomass was found reduced by ca. 39 per cent as compared to control. When (IAA + UV-B), (Kn + UV-B) and (GA_3 + UV-B), was considered, the total biomass was found to improved by ca. 32.4 per cent, 31.4 per cent and 30.8 per cent as compared to UV-B treatment alone.

11. Net primary production was calculated at 15th day intervals of growth for both the varieties of mustard crops in all treatments. Overall, results of net production follow the trend of standing crop of biomass. UV-B was found to reduce the production potential when given individually and an improvement was recorded when subjected along with PGRs over individual treatment of UV-B. The total net production was found to reduce was noted at 105th day stage of growth. When the crop was treated with IAA along with UV-B radiation, the maximum value of T.N.P. were recorded at 105th day stage of growth and improved by ca. 17.9 per cent as compared to individual UV-B treatment in case of *Brassica compestris PT-303.* When the crop was treated with Kn and $GA_{3,}$ along with UV-B radiation, the maximum value of T.N.P. were recorded at 105th day stage of growth and improved by ca. 25.2 per cent and 26.7 per cent as compared to individual UV-B treatment.

12. In case of *Brassica juncea PR-15,* the crop was treated with UV-B treatment along with IAA, the maximum value of T.N.P. were found at 105th day stage and promoted by 12.7 per cent as compared to individual treatment of UV-B. It was found increased when treated with combination of PGRs and the maximum enhancement was recorded by 25.2 per cent with (Kn + UV-B) and 25.3 per cent with (GA_3 + UV-B) as compared to UV-B treatment only.

13. Storage of dry matter was calculated for both varieties of mustard crops as affected by UV-B individually and in combination with PGRs on the basis of average biomass. In case of *Brassica compestris PT-303,* dry matter was found affected by ca. 30 per cent decrease in live matter, 33 per cent decrease in above ground dry matter and 25 per cent decrease in under ground. The different above ground plant parts *viz.* stem, leaves, flower and fruits etc. were accounted a decrease of ca. 42 per cent, 26 per cent, 10 per cent, 29 per cent respectively as compared to control. The maximum ameliorative effect against UV-B of storage of dry matter was found with UV-B + IAA amounted ca. 23 per cent for live vegetation, 28 per cent for above ground and 16.3 per cent for underground as compared to UV-B only. In Kn + UV-B amounted

ca. 24.3 per cent for live vegetation, 30 per cent for above ground and 16.5 per cent for underground as compared to UV-B only. In GA_3 + UV-B amounted ca. 24.5 per cent for live matter, 30 per cent for above ground and 16.5 per cent for underground matter respectively as compared to individual treatment of UV-B exposure.

When *Brassica juncea PR-15* was treated with UV-B only, a total decrease of ca. 25 per cent for live vegetation, 27 per cent for above ground, 22 per cent for underground and 26.7 per cent for stem, 26.5 per cent for leaves and 25.9 per cent for flower and 26.6 per cent for fruits respectively as compared to control. When the crop was treated with UV-B along with IAA, the promotion was noted and amounted as ca. 18.9 per cent for live vegetation, 21.9 per cent for above ground, 13.8 per cent for under ground, 25.3 per cent for stem, 17.8 per cent for leaves, 12.9 per cent for flower and 22.4 per cent for fruits respectively as compared to individual treatment of UV-B. When crop was studied with Kn + UV-B radiation, it showed as ca. 19.7 per cent, 24.4 per cent, 13.8 per cent, 30.3 per cent, 17.6 per cent, 14.9 per cent and 21.7 per cent for live vegetation, above ground, under ground, stem, leaves, flower and fruits. When crop was studied with GA_3 + UV–B radiation, it showed as ca. 18.6 per cent, 21.9 per cent, 14.2 per cent, 25.4 per cent, 17.6 per cent, 15.9 per cent and 21.4 per cent for live vegetation, above ground, under ground, stem, leaves, flower and fruits respectively as compared to UV-B exposure only.

14. The dry matter dynamics was calculated on the basis of net primary productivity of different plant parts as affected by various treatments in both the varieties of mustard crops. A decrease of ca. 29 per cent and 24 per cent was recorded in total net primary productivity due to individual treatment of UV-B in *Brassica compestris PT-303* and *Brassica juncea PR-15* respectively. An improvement in T.N.P. was reported due to combined treatment of UV-B and IAA and it was increased by ca. 23.2 per cent and ca. 17.8 per cent for *Brassica compestris PT-303* and *Brassica juncea PR-15* respectively.

15. Harvesting index, a measure of economic yield in cropland ecosystem, was also studied in present study under influence of different treatments. A decrease of ca. 4 per cent and ca. 4.3 per cent was observed in *Brassica compestris PT-303* and *Brassica juncea PR-15 respectively* due to UV-B radiation. IAA was found to be most counteracting when given along with UV-B radiation and improved by ca. 2.5 per cent and 2.4 per cent in *Brassica compestris PT-303* and *Brassica juncea PR-15* respectively. These results showed that *Brassica compestris PT-303* responded more significantly than *Brassica juncea PR-15* in combined treatment of UV-B and IAA.

16. Shelling Percentage of *Brassica compestris PT-303* and *Brassica juncea PR-15* were studied and decreased by ca. 5 per cent and 3 per cent respectively by individual treatment of UV-B exposure as compared to control. When UV-B exposure was given along with IAA and Kn, the maximum mitigation was noticed and increased by ca. 5 per cent and 6 per cent respectively as

compared to individual treatment of UV-B exposure in *Brassica compestris PT-303* and *Brassica juncea PR-15* respectively.

17. Experimental studies showed pronounced effect of UV-B exposure and PGRs individually and in combination on chlorophyll development of seedlings of both the varieties of mustard crops. Results of the present study showed decrease of ca. 23 per cent, 17 per cent and 6 per cent of chlorophyll a, chlorophyll b and protochlorophyll respectively in case of *Brassica compestris PT-303* and ca. 16 per cent, 24 per cent 6 per cent for chlorophyll a, chlorophyll b, protochlorophyll in case *Brassica juncea PR-15* respectively after 7th days of seedling growth. When the seedlings were treated with UV-B along with IAA, it showed as 14 per cent, 21.1 per cent and 10 per cent for chlorophyll a, chlorophyll b and protochlorophyll in case of *Brassica compestris PT-303*. In case of *Brassica juncea PR-15,* the crop was treated with UV-B along with IAA and it showed as 2.6 per cent, 10 per cent, 4.3 per cent for chlorophyll a, chlorophyll b, protochlorophyll. Kn + UV-B showed as 16.6 per cent, 2.2 per cent, 4.6 per cent for chlorophyll a, chlorophylll b, protochlorophyll respectively. GA_3 + UV-B showed as 2.7 per cent, 13.4 per cent, 6.4 per cent for chlorophyll a, chlorophylll b, protochlorophyll respectively.

    When individual treatment of UV-B was given to field grown crops a decline of 20 per cent, 17 per cent and 18 per cent in case of *Brassica compestris PT-303* and 26 per cent, 29 per cent and 23 per cent in *Brassica juncea PR-15* was recorded for chlorophyll a, chlorophyll b and protochlorophyll respectively. When these crops were supplemented with combination of UV-B and PGRs, an increase in different chlorophyll contents was recorded. Out of the two PGRs, was found to an improved the chlorophyll contents by 19 per cent, 45 per cent and 5 per cent for chlorophyll a, chlorophyll b and protochlorophyll respectively for (IAA + UV-B), 15 per cent, 10 per cent and 7 per cent of chlorophyll a, chlorophyll b and protochlorophyll (GA_3 + UV-B) in case of *Brassica compestris PT-303*. In case of *Brassica juncea PR-15,* PGRs were treated with UV-B radiation. IAA was observed maximum promotory for different chlorophyll pigment as compared to UV-B treatment alone. The maximum promotion in chlorophyll a, chlorophyll b and proto chlorophyll was noted as 14 per cent, 8 per cent and 9 per cent respectively.

18. Plants develop anthocyanin as a protection pigment against UV-B radiation as evidenced by present as well as other experimental studies. When the seedlings were treated with UV-B, a marked promotion of ca. 128 per cent and ca. 153 per cent of anthocyanin pigment was recorded due to UV-B treatment in *Brassica compestris PT-303* and *Brassica juncea PR-15* respectively. IAA, Kn and GA_3 was found to mitigate the effect of UV-B radiation and consequently lowers down the accumulation of pigment as compared to UV-B treatment in *Brassica compestris PT-303.*

19. The protease activity was also enhanced in the germinating seeds of both varieties of mustard crops due to UV-B exposure. A rise of ca.115 per cent after 6 hrs; 124 per cent after 12 hrs. and 120 per cent after 24 hrs. respectively

in *Brassica compestris PT*-303 as compared to control. In *Brassica juncea PR-15,* a rise of ca.114 per cent, 122 per cent and 139 per cent was noticed after 6 hrs; 12 hrs. and 24 hrs. respectively as compared to control. When the seeds were supplied with PGRs *viz.* IAA, Kn and GA_3 along with UV-B treatment which changed significantly and a decrease of ca. 11 per cent, 15 per cent, 19 per cent; 22 per cent, 17 per cent, 2 per cent; 40 per cent, 13 per cent 3 per cent at 6 hrs, 12 hrs. and 24 hrs. reported in *Brassica compestris PT-303* and *Brassica juncea PR-15* respectively as compared to UV-B exposure alone.

20. A marked increase in peroxidase was noted in inhibited seeds of both the crops due to UV-B radiation. An increase of ca. 68 per cent, 123 per cent, 108 per cent was recorded in *Brassica compestris PT-303* and ca. 105 per cent, 113 per cent, 106 per cent in *Brassica juncea PR-15* after 6 hr, 12 hr and 24 hr of imbibitions respectively due to UV-B (3 hrs. daily) radiation as compared to control. When the seeds were supplied with PGRs along with the above treatment, this effect was altered significantly and a decrease of ca. 3 per cent, 41 per cent and 24 per cent with IAA at 6 hr, 12 hrs and 24 hrs; with Kn ca. 2 per cent, 12 per cent, 37 per cent at 6 hrs, 12 hrs and 24 hrs and with GA_3 ca. 42 per cent, 42 per cent, 24 per cent at 6 hrs, 12 hrs and 24 hrs. respectively was reported in *Brassica compestris PT-303*. In case of *Brassica juncea PR-15,* a decrease of ca. 7 per cent, 23 per cent, 21 per cent with IAA; ca. 7 per cent, 30 per cent, 36 per cent with Kn and ca. 9 per cent, 34 per cent, 44 per cent with GA_3 at the 6 hrs 12 hrs and 24 hrs respectively as compared to UV-B exposure only.

All the parameters considered during the present study such as shoot length, leaf area, fresh weight and dry weight of leaf, stem, root, flower, TNP, harvesting index, shelling percentage etc. were found to be decreased when exposed to UV-B radiation. Photosynthetic pigments *viz.* chlorophyll a, chlorophyll b and protochlorophyll and enzymes protease as well as peroxidase were also reduced when subjected to UV-B radiation. Effect of UV-B on Brown sarson and Rai, as far as anthocyanin is concerned was reported an enhancement. It can be assumed after overall studies that accumulation of anthocyanin because of UV-B could act as a screen by absorbing UV-B radiation and in turn protect the chloroplast from UV-B induced damage.

When these most important cereals *viz.* Brown sarson and Rai were treated with UV-B along with PGRs (IAA, Kn and GA_3), a counteracting effect was reported in all the parameters studied. So, it has been concluded in our study that these plant growth regulators (IAA, Kn and GA_3) can mitigate the hazardous or deleterious effects caused by UV-B in these varieties of mustard crops significantly.

References

Agarwal, M., Agarwal S.B., Krizek, D.T., Krammer, G.P. Lee Edward and Rowland, R. A. (1991). Physiological and morphological response of snapbean to ozone stress and pretreatment with UV-B radiation in: impact of global climate change on photosynthesis and plant productivity. Y.P. Abrol *et al.* Oxford and IBH publishing co. Ltd. India. PP, 133-148.

Agarwal, S.B., D. Rathore S.A. Singh. (2004). Combined effects of enhanced UV-B radiation and additional nutrients on two cultivars of wheat (*Triticum aestivum* L). *Physiology and molecular Biology of plants* **10:** 99-108.

Alexieva, V., I. Sergiev, S. Mapelli and E. Karanov. (2001).The effect of drought and ultraviolet radiation on growth and stress markers in pea and wheat. *Plant, cell and Environment* **24:** 1337-1344.

Allen, D.J, McKee I.F, Farage P.K, Baker NR. (1997). Analysis of the limitation to CO_2 assimilation on exposure of leaves of two *Brassica napus* cultivars to UV-B. *Plant, cell and Environment* **20:** 633-40

Ambasht, N.K. and M. Agarwal. (1995). Physiological responses of field grown *Zea mays* L. Plants to enhanced UV-B radiation. *International journal of Biotronics* **24:**15-23.

Ambler, J.E., R.A. Rowland and N.K. Maher. (1978). Response of selected vegetable and agronomic crops to increase UV-B irradiation under field conditions. In UV-B biological and climatic effects research (BACER). Final Report.

Ambrish, K. (1992). Effect of supplemental UV-B radiation on growth and composition of certain Legume crops. D. Phil thesis, H.N.B. Garhwal University Srinagar (Garhwal).

Andreia Henrique, Eduardo Nogueira Campinhos, Elizabeth Orika Ono, Sheila Zambello D. E. Pinho. (2006). Effect of Plant Growth regulators in the rooting of Pinus cuttings. Braz. Arch. Boil. Technol. **49** (2) Curitiba Mar.

Andreae, W. A. and S. R. Andreae. (1953). Studies on indole acetic acid metabolism. I. The effect of methyl umbellifeone, maleic hydrazide and 2,4–D on indole acetic acid oxidation. *Can. J. Bot.* **31**: 426–437.

Antognozzi E, Battistelli A, Famiani F, Moscatello S, Stanica F, Tombesi A. (1996). Influence of CPPU on carbohydrate accumulation and metabolism in fruits of *Actinidia deliciosa* (A. Chec.).*Sci Hort.* **65** :37-47.

Arakawa, O., Y. Hori and R. Ogata. (1985). Relative effectiveness and interaction of ultraviolet–B, red and blue light in anthocyanin synthesis of apple fruit. *Physiol. Pl.* **64**: 323–327.

Arney, S.E. and P. A. Mancinelli. (1967). Surgical investigations of the role of IAA and GA in elongation of "Meteor" pea internodes. *New phytol.* **66:** 271-283.

Arya, D., Patni, V. (2007). Effect of different Plant Growth regulators on multiple shoot induction in *Pluchea lanceolata* Oliver and Hiern J of *Phytological Research* **20** (1): 29-32.

Awasthi, Anjali, Uniyal, Sanjay Kumar, Rawat Gopal, S. (2001). Forest management down the ages: A case study from District Uttarkashi (Uttarakhand). *Indian journal of forestry*, Vol. **24 (3):** 388-394.

Badawi, M.A. and Sahhar, E.L. (1978). Influence of some growth substances on different characters of cabbage. *Egyptian J. Hort.*, **6** (2); 221-285.

Bahuguna, V.K., Dhawan, V.K., and Pant, B.D. (1988). Studies on the effects of Growth Hormones for vegetative propagation of *Woodfordia fructicosa Kurz.* By rooting of branch cuttings. *Indian Forester*, **141:** 832-836.

Baker, N.R, Nogues S, Allen, D.J. (1997). Photosynthesis and photoinhibition. In: Lumsden PJ, ed. *Plants and UV-B responses to environmental change.* Cambridge University Press, 95-111.

Baker, N.R, Bowyer, J.R. (1994). *Photoinhibition of photosynthesis: From molecular mechanisms to the field.* Oxford: Bios. Scientific publishers.

Ballare, C.L., P.W. Barnes and S.D. Flint. (1995). Inhibition of hypocotyl elongation by UV-B radiation in de-etiolating. The photoreceptor. *Physiologia plantarum* **93:** 584-592.

Ballare, Searles P.S, Robson, T.M, Sala O. E, Scopel A.L. (2001). Impacts of solar UV-B radiation on the terrestrial ecosystems of Tierra del Fuego (Southern Argentina). *J. Photochem. Photobiolo.* **62:** 67-77.

Baraldi R, F. Rossi and B. Lercari. (1988). *In vitro* shoot development of *prunus* GF6652: interaction between light and benzyladenine. *Plant Physiol.* **74:** 440-443.

Barnes PW, Flint S.D, Caldwell M.M. (1990). Morphological responses of crop and weed species of different growth forms to UV-B radiation. *Am. J. Bot.* **77:** 1354-1360.

Barnes, B.V., D.R. Zak, S.R. Denton and S.H. Spurr, (1998). Forest ecology, 4th edition. John Wiley and Sons. New York, 774p.

Barsing, M. and R. Malz. (2000). Fine structure, carbohydrates and photosynthetic pigments of sugar maize leaves under UV-B radiation. *Environmental and Experimental Botany* **43:** 121-130.

Bassman, J.H, G.E. Edwards and R. Robberecht. (2003). Photosynthesis and growth in seedlings of five forest tree species with contrasting leaf anatomy subjected to supplemental UV-B radiation. *For sci.* **49:** 176-187.

Bennett, J. H. (1981). Photosynthesis and gas diffusion in leaves of selected crop plants exposed to ultraviolet–B radiation. *J. Environ. Qual.* **10**: 271–275.

Bieza, K. and R. Lois. (2001). An *Arabidopsis* mutant tolerant to lethal UV-B levels shows constitutively elevated accumulation of flavonoids and other phenolics. *Plant physiol.* **126:** 105-115.

Biggs, R.H., S.V Kossuth. (1978). Effects of UV-B radiation enhancement under field conditions. In UV-B Biological and climatic Effects Research final Report.

Biggs, R.H., S.V. Kossuth and A.H. Teramura. (1981). Response of 19 cultivars of soybeans to UV-B irradiance. *Physiol. Pl.* **53:** 19-26.

Bisht, N.S. (1981). Studies on community structure, organic productivity and energetics in Grass-land at Kotdwara. D. Phil thesis, H.N.B. Garhwal University. Srinagar (Garhwal).

Blackman, P.G. and Davies, W. J. (1985). Root to shoot communication in maize plants of the effects of soil drying. *J. Exp. Bot.* **36:** 39-48.

Boccalandro, H.E, Mazola M.A, Ballare C.L. (2001). UV-B radiation enhances a phytochrome-B mediated photomorphogenic response in *Arabidopsis plant physiol.* **126:** 780-788.

Bosabalidis, A.M. and Skoula, M. (1998). A comparative study of the glandular trichomes on the upper and lower leaf surface of origamum X intercedens Rech. *Journal of Essential Oil Research* **10:** 277-286.

Brian, P.W., and H.G. Hemming. (1958). Complementary action of gibberellic acid and auxins in pea internodes extension. *Ann. Bot.* **22:** 1-17.

Buckovac, M.J. and Wittwer, S.H. (1957). GA and higher plants, III induction of flowering in biennials. Quart Bull. Mich. *Agric expt. Sta.,* **39:** 650-660.

Caldwell, M.M, Ballare, C.L, Flint, S.D, Bjorn, L.O. Teramura, A.H, Kulandaivelu G. and Tevini, M. (2003). Terrestrial ecosystems increased solar UV radiation and interactions with other climatic change factors, *Photochem. Photobiolo. Sci.* **2:** 29-38.

Canomedrano R. and Darnell, R.L. (1997). Sucrose metabolism and fruit growth in parthenocarpic vs. seeded blueberry (*Vacinium ashei*) fruits. *Plant. physiol.* **99:** 439-446.

Cen Y-P, Bornman, J.F. (1993). The effects of exposure to enhanced UV-B radiation on the penetration of monochromatic and polychromatic UV-B radiation in leaves of *Brassica napus. Physiol.* **87:** 249-255.

Chaudhary, B.L. and Vini Vyas. (2002). Effect of IAA on spore germination in *Funaria* Hygrometrica Hedw. *J. Indian Bot. Soc.* **85:** 135-137.

Chauhan, K.S. and Balwant Singh. (1970). Response of cabbage to foliar applications of Gibberellic acid and urea. *Indian J. Hort.* **27:** 68-71.

Chhonkar, V.S. and Jha, R.N. (1963). The use of plant growth regulators in transplanting of cabbage and their response on growth and yield. *Indian. Hort.* **20:** 123-128.

Chibnall, A.C. (1954). Protein Metabolism in rooted runner bean leaves. *New Pytol.* **53:** 31-37.

Chimpango, S.B.M., C.F. Musil and F.D. Dakora. (2003). Effects of UV-B radiation on plant growth, symbiotic function and concentration of metabolites in three tropical grain legumes. *Funct. Plant Biol.* **30:** 309-318.

Correia, C.M., J.F. Coutinho, L.O. Bjorn and J.M.G. Torres-pererira (2000). UV-radiation and nitrogen effects on growth and yield of maize under Mediterranean field conditions. *Eur. J. Agron.* **12:** 117-125.

Cuadra, P.R. Herrera and V. Fajardo. (2004). Effects of UV- B radiation on the Patagonian (*Jaborosa magelenica*) Brisben. *J. Photochem. Photobilo.B.* **76:** 61-68.

Day, T.A. and Vogelmann, T.C. (1995). Alterations in photosynthesis and pigment distributions in pea leaves following UV-B exposure. *Physiologia Plantarum* **94:** 433-440.

Demchik, S.M. and T.A. Day, (1996). Effects of enhanced UV-B radiation on pollen quantity, quality and seed yield in *Brassica rapa. American journal of Botany* **93:** 573-579.

Dhasmana, R. (1984). Effects of Growth Regulators on the productivity, Energy budget and Mineral cycling of *Medicago sativa.* D. Phil thesis. Garhwal University, Srinagar (Garhwal).

Dickson, J. G. and M. M. Caldwell, (1978). Leaf development of *Rumex patientia* L. (*Polygonaceae*) exposed to UV irradiation (280–320 nm). *Amer. J. Bot.* **65** (8): 857–863.

Dillenburg, L.R., J.H. Sullivan and A.H. Teramura, (1995). Leaf expansion and development of photosynthetic capacity and pigments in *Liquidamber styraciflue* (Hamamelidaceae). Effects of UV-B radiation. *American Journal of Botany.* **82:** 878-885.

Dunning, C.A., L. Chalker-Scott and J.D. Scott, (1994). Exposure to UV-B radiation increases cold hardiness in *Rhododendron. Physiologia Plantarum* **92:** 516-520.

Duysen, M., K. Eskins and L. Dybas, (1985). Blue and white light effects on chloroplast development in a soybean mutant. *Photochem. Photobiol.* **41**: 667 – 672.

Elena, K.A. Wulff and R. Julkunen-Titto. (2001). Growth, structure, stomatal response and secondary compounds of birch seedlings (*Betula pendula*) under elevated UV-B radiation in the field *Tree. Physiol.* **18:** 53-58.

Fan Wu, Jen-Tsung Chen and Weichin chang. (2004). Effects of Auxins and Cytokinin on Embryo formation form root- derived callus of Oncidium 'Gower Remsey'. Plant Cell, Tissue and Organ Culture. **77**(1): 107-109.

Feng, H.L., A.N., Chen. W. Qiang. S. XU. M. Xiao, X. Wang and G. Cheng, (2003).The effects of enhanced UV-B radiation on growth, photosynthesis and stable carbon isotope composition of two soybean cultivars (*Glycine max*) under field conditions. *Environ Exp. Bot.* **49:** 1-8.

Fiscus, E.L. and F.L. Booker, (1995). Is increased UV-B threat to crop photosynthesis and productivity? *Photosyn. Res.,* **43:** 81-92.

Francis, C.A. Flor and M. Prager. (1978). Effects of bean association on yield and yield components of maize. *Crop Sci.* **18** (5): 760-764.

Gala, W.R. and J.P. Giesy, (1991). Effects of UV-B radiation on primary productivity of natural phytoplankton assemblages in Lake Michigan. Ecotoxicol. *Environ. Safety,* **22:** 345.

Gehrke, C.U. Johnson, T.V. Callaghan. D. Chadwick and C. Robinson, (1995). The impact of enhanced UV-B radiation on litter quality and decomposition processes in *Vaccinium* leaves from the Subarctic Oikos. **72:** 213-222.

Gilford, J.M.D.& Rees, A.R. (1973). Growth of the tulip shoot. *Sci. Hort.* **1:** 143-156.

Gonge, V.S., Birjees Hameed, S.G. Bharad, A.D. Warade and S.N. Kale. (2005). Effect to plant growth regulators on yield and quality of Okra Seed. PKV Res. J. **29** (1): 7-9.

Gonzalez R, Paul N.D, Percy K, Ambrose M, McLaughlin C.K, Barnes J.D, Areses M. and Welburn A.R. (1998). Responses to UV-B radiation (280-315 nm) of pea (*Pisum sativum*) lines differing in leaf surface wax. *Physiologia plantarum* **98:** 852-860.

Goyal, A.K. and V.K. Jain, (1990). UV-induced inhibition in growth and productivity of Linseed crop. *Symposium on photo-physiology and photomedicine,* New Delhi.

Goyal, A.K. and V.K. Jain and K. Ambrish, (1991). Effect of supplementary UV-B radiation on the growth, productivity and chlorophyll of field grown Linseed crop. *Indian J. Pl. Physiol.* **34** (2): 374-377.

Green and Neurath. (1954). The proteins. Neurath, H. and K. Bailey K (ed.) 2 B Academic Press, New York.

Greene, O. (1995). Emerging challenges for the Montréal protocol. The Globe **27:** 5-6.

Gutkin, S.S. (1950). Properties and uses of high iodine number falkidine drying oils, *J. Amer. oil chem. Soc.* **27:** 542.

Gzik, A., Schonfield, G., Baumann, I. and Gunther, G. (1987). Interactions of different phytohormones on photosynthetic pigment content and nitrate reeducates activity in leaf discs of sugar beet. pp. 547-551. *Sofia Bulgaria. M. Popov. Inst. PI. Physiol.*

Halvey, A.H. and Mayak S. (1981). Senescence and post harvest physiology of cut flowers. *Part II. Hotric. Rev.* **3:** 59-193.

Han, S.S. (2001). Benzyladenine and gibberellins improve post harvest quality of cut Asiatic and oriental lilies. *Hort-Science* **36 (4):** 741-754.

Hanks, G.R. (1982). The response of tulips to gibberellins following different durations of cold storage *J. Hortc. Sci.* **57:** 109 119.

Hart, R.H., G.E. Carlon, H.H. Klueter and H.R. Carns. (1975). Response of economically valuable species to ultraviolet radiation. In climatic impact assessment program. (CIAP) **5:** 263-275.

Hashimoto, T., C. Shichijo and H. Yatsuhashi. (1991). Ultraviolet action spectra for the induction and inhibition of anthocyanin synthesis in Broom *Sorghum* seedlings. *Photochem. Photobiol.* **11** (3): 353–364.

Hayada Y, Nimi Y, Iwasaki N. (1958). Inducing Parthenocarpic fruit of watermelon with plant bioregulators. *Acta Hort* **394:** 235-240.

Heinrich, G.H.K., C. Schmude, H. Garden, O.Y. Koroleva and K. winter (1999). Effects of solar UV-B radiation on the potential efficiency of Photosystem-II in leaves of tropical plants. *Plant physiol.* **121:** 1349- 1358.

Hidema, J., T. Kumagai and B.M. Sutherland, (2000). UV-radiation sensitive Norin-1 Rice contains deftective cyclobutane pyrimidine dimmer photolyase *Plant cell* **12:** 1569-1578.

Hofmann, R.W., B.D. Campbell, D.W. Fountain, B.R. Jordan, D.H. Greer, D.Y. Hunt and C.L Hunt. (2001). Multivariate analysis of intraspecific responses to UV-B radiation in white clover (*Trifloium repens* L). *Plant cell environ.* **24:** 917-927.

Holm, R.E. and J.L. Key, (1969). Hormonal regulation of cell elongation in the hypocotyl of rootless soybean: An evaluation of the role of DNA synthesis. *Pl. Physiol.* **44:** 1295-1302.

Hong-Yi Li, Kai-Wen Pan, Qing Liu and Jin Chuwang. (2008). Chengdu Institute of Biology, Chinese Academy of Sciences, (China).

Hossain, Z. Mandal, A.K.A, Datta, S. K. and Biswas, A.K. (2006). Decline in ascorbate peroxidase activity-a prerequisite factor for tepal senescence in gladiolus. *J. Plant physiol.* **163 (2):** 186-194.

Hunt, J.E. and D.L. McNeil, (1998). Nitrogen status affects UV-B sensitivity of cucumber. *Australian Journal of plant physiology.* **25:** 79-86.

Irullappan, I. and C. R. Muthukrishnan, (1973). Effect of growth regulators on the plant composition and nutrient uptake of L. *Esculentum* Mill. Madras *Agric.* **J. 60** (9/12): 1635-1643.

J.T. Baker, L.H. Allen J.R. (1994). Assessment of the impact of rising CO_2 and other potential changes on vegetation. *Environ. Pollu.* **3:** 223-235.

Jain, Vanita, Meena R.C. (1999). Changes in growth and photosynthesis of mungbean induced by UV-B radiation. *Indian journal. Pla. Phy.* Vol. **4pp** 79-84.

Jain, V. K. and A. K. Goyal, (1990). Chlorophyll development response to supplemented UV–B in Lentil crop under field conditions. Symposium on Photophysiology and Photomedicine, New Delhi.

Jansen, M.A.K, Gaba V, and Greenberg, B.M. (1998). Higher plant and UV-B radiation: balancing damage, repair and acclimation. *Trends plant Sci.* **3:** 131-135.

Jauhari, O. S., R. D. Singh and V. S. Dikshit. (1960). Preliminary studies on the effect of gibberellic acid on growth of spinach (*Spinacia oleracea*). *Curr. Sci.* **29**: 484–485.

Jawanda, J.S., Josan, J.S. and singh, S.N. (1979). Propagation of Prunus Species by cuttings. I. Effect of IBA and the type of Peach cutting on rooting. *Journal of Research, India* **16:** 408-412.

Jones, R.L, and I.D.J. Phillips. (1966). Organs of gibberellins synthesis in light grown sun-flower plants. *Plant physiol.* **43:**1381-1386.

Jordan, W. R. and F. Skoog. (1971). Effects of cytokinin on growth and auxin in coleoptiles of derooted Avena seedlings. *Pl. Physiol.* **48**: 97-99.

Jordan, B.R., He, J., Chow, W.S. and Anderson, J.M. (1992). Changes in RNA levels and polypeptide subunits of RUBP in response to supplementary UV-B. *Plant cell Enviorn.***15:** 91-98.

Jordon, B.R. (1996). The effects of ultraviolet-B radiation on plants: a molecular perspective. *Advances in Botanical Research* **22:** 97-161.

Kakani, V.G., K.R. Reddy, D. Zhao and K. Sailaja. (2003). Field crop response to UV-B radiation: a review. *Agric. For. Meterol.* **120:** 191-218.

Karou R, Grammatikopulos G, Manetas Y, Kokkini S. (1998). Effects of UV-B radiation on *Mentha spicata* essential oil. *Phytochemistry* **49:** 2273-2277.

Kaufman, P.B. Ghesheh, N. Ikuma, H. (1968). Promotion of growth and invertase activity by gibberellic acid in developing *Avena* internodes. *Plant physiol.* **43:**29-34.

Kaur, S., A.K. Gupta and N. Kaur, (1998). Gibberellic acid and kinetin partially reverse the effect of water stress on germination and seedling growth in chickpea. *Pl. Growth Regul.* **25:** 29-33.

Keiller, D.R. and M.G. Holmes, (2001). Effects of long term exposure to elevated UV-B radiation on the photosynthetic performance of five broad leaves tree-species. *Photosynth. Res.* **67:** 229-240.

Key, J.L. and J.C. Shanon, (1964). Enhancement by auxin of nucleic acid in excised soybean hypocotyl tissue. *Pl physiol.* **39:** 360-364.

Khalil, S. and Mandurahi, H.M. (1989). Combined effect of soil water availability and growth substances on some aspects of chemical composition of cowpea plants. *J. Agrun, Crop. Sci.* **162:** 81-92.

Khan, A.A and K.L. Tao, (1978). Phytohormones, seed dormancy and germination. In: Phytohormones and Related compounds. A comprehensive Treatise. 371-422. Vol. 2 D. Letham, P.B. Goodwin and T.J.V. Higgins (eds.) Elsevier/North-Holland Biochemical Press, Amsterdam.

Kim, B.C., Tennessen, D.J., Last. R.L. (1998): UV-B induced photomorphogenesis in *Arabidopsis thaliana. Plant J.* **15:** 667-674.

Kim, H.Y., K. Kobayashi, I. Nouchi and T. Yoneyama. (1996). Differential influences of UV-B radiation on antioxidant and related enzymes between rice (*Oryza sativa*) and cucumber (*Cucumis sativas* L) leaves. *Environ. Sci.* **9:** 55-63.

Kolb, C.A., M.A. Kaser, J. Kopecky, G. Zotz, M. Reader and E.E. Pfounded. (2001). Effects of the natural intensities of visible and UV- radiation on epidermal ultraviolet screening and photosynthesis in grape leaves. *Plant physiol.* **127:** 863-875.

Koski, V.M. and J.H.C. Smith, (1948). The isolation and spectral observation properties of protochlorophyll from etiolated barley seedlings. *J. Amer. Chem. Soc.* **70:** 35-58.

Krishnamurthy, L. (1986). Studies on growth, development and source sink relationship in mustard *(Brassica junea* L. Ph.D. Thesis, Institute of Agricultural sciences, B.H.U. Varanasi.

Krishnamoorthy, H. N. (1992). Effects of the growth retardants and abscisic acid on the rooting of hypocotyl cuttings of musk melon (*Cucumis melo* CV Kutana). Biochem. *Physiol. Pflanzen.* **163**: 513-517.

Krizek, D.T., Britz, S.J., Mireki, R.M. (1998). Inhibitory effects of ambient levels of solar UV-B radiation on growth of CV. New Red Fire lettuce. *Plant physiol.* **130:** 1-7.

Krupa, S.V. and Kickert, R.N. (1989). The green house effect: Impacts of UV-B radiation, CO_2 and ozone (O_3) on vegetation. *Environmental pollution.* **61:** 263-393.

Kulandaivelu G, Nedunchezhian N. (1993). Synergistic effects of ultraviolet-B, enhanced radiation and growth temperature on ribulose 1,5- bisphosphate and $^{14}CO_2$ fixation in *Vigna sinensis L. Photosynthetica* **29:** 377-383.

Kulandaivelu, G. and K. Annamalainathan, (1991). Interaction of herbicide and UV-B radiation on the photosynthetic apparatus pp59-75. In: Y. P. Abrol, P.N. Wattal, A. Gnanam, ort. D.R. Govindjee and A.H. Teramura. Impacts of Global climatic changes on photosynthesis and plant productivity. Oxford and IBH publication, New Delhi.

Kumar N. Srivastava G.C. and Dixit K. (2007). Role of superoxiadse dismutase (SOD) during petal senescence in rose *(Rosa hybrida L.) J. Hortic Sci Biotechnol* **82 (5):** 673-678.

Kumar, A. (1981). Effect of Growth regulators on growth pattern, productivity, mineral cycling and energy budget of groundnut (*Arachis hypogaea* L.). D. Phill. Thesis submitted to Meerut University (Meerut).

Kumar, D.K.D., Paliwar, R. and Kumar, D. (1996). Yield and yield attributes of cabbage as influenced by GA and NAA. *Crop Res. Hisar,* **12 (1):** 120-122.

Kumar, R., M. M. Sharma, V.K. Jain and A.K. Goyal (1988). Growth of response of Lentil crop to UV-B irradiation under field condition. *Indian Jour. Pl. Physiol*, **31:** 297-300.

Kumari, S. and Bharti, S, (1992). Effect of CCC and FAP on photosynthesis in sunflower under simulated drought condition. *Haryana Agri. Univ. J. Res.* **22:** 206-213.

Kuraichi, S. and R. M. Muir. (1963). Diffusable auxin increase in a rosette plant treated with gibberellin. *Naturewiss 50*: 337–338.

Laakso, K. and S. Huttuncn, (1998). Effects of UV-B radiation on conifers: a review. *Environ. Pollu* **99:** 319-328.

Lander and Morisson, (1962). The Wealth of India. A dictionary of Indian Raw Materials and industrial products. CSIR, India **6:** 120-132.

Lavola, A. (1998). Accumulation of flavonoids and related compounds in birch induced by UV-B irradiance. *Tree physiol.* **18:** 53-58.

Levizou, E and Y. Manetas, (2001). Combined effects of enhanced UV-B radiation and additional nutrients on growth of two Mediterranean plant species. *Plant Ecology* **154:** 181-186.

Liakoura V, Manetas Y, Karabourniotis G. (2001). Seasonal fluctuations in the concentration of UV-B absorbing compounds in the leaves of some Mditerranean plants under field conditions. *Physiologia Plantarum* **111:** 491-500.

Lockhart, J.A. (1964). Physiological studies on the light sensitive stem growth. *Planta* **62:** 97-115.

Madronich S, McKenzie R.L. Caldwell M.M. and Bjorn L.O. (1995). Changes in UV-radiation reaching the earth's surface. *Ambio.* **24**: 143-52.

Maffei M, Scannerini S. (2000). UV-B effects on photomorphogenesis and essential oil composition in peppermint (*Mentha piperita*). *Journal of essential oil Research* **12:** 523-529.

Mahely and Chance. (1967). The assay of catalases and peroxidase. 357-423. In; Methods of Biochemical Analysis. Ed. Glick Inter Science publication Inc., New York.

Mahmud, B.S. (1983). The Effects of two Chemical Growth Regulators on Oilseed Rape. Ph.D. Thesis, University of London, Wye College, Ashford.

Mancinelli, A.L., C.P.H., Yang, P. Lidquist, O.R. Anderson and I. Rabino, (1975). Photocontrol of anthocyanin synthesis III. The action of streptomycin on the synthesis of chlorophyll and anthocyanin. *Pl. Physiol.* **55:**251-257.

Mazza, C.A, D. Battista, A.M. Zima, M. Szwarcberg Bracchitta, C.V. Giodano, A. Acevedo, A.L. Scopel and C.L. Ballare, (1999). The effects of UV-B radiation on the increased DNA damage and antioxidant responses Plant cell Environ. **22:** 61-70.

McKenzie R.L, Bjorn.L.O. and Bais A, (2003). Changes in biologically active UV-radiation reaching the earth's surface. *Photochem. Photobilo. Sci.* **2:** 5-15.

Michael Barsing and Ralf Malz, (2000). Fine structure, carbohydrates and photosynthetic pigments of sugar maize leaves under UV-B radiation. *Environ. And exp. Botany* **43:** 121-130.

Mishra, R.S., Tewari, J.P. and Joshi, K.R. (1986). Effect of IBA, Boron and Catechol of the rooting of commercial cultivars of Plum Grown in U.P. Hills. *Prog. Hort.* **18:** 24-28.

Miyamoto K, Ueda J. Kamisaka S. (1993). Gibberellin enhanced sugar accumulation in growing subhooks of etiolated *Pisum sativum* seedlings. Effect of gibberelic acid, IAA and cycloheximide on invertase activity, sugar accumulation and growth. *Plant physiol.* **88:** 301-306.

Moe, R. and A.S Anderson, (1987): Effects of stock plant environment on subsequent adventitious rooting of cuttings. In: Adventitious Root Formation on cuttings. T.D. Davis *et al.* (eds).

Mohariya, A.D., Patil, B.N., Wankhede, S.G., Band, P.E. and Kartikayan, Reena, (2003). Effect of GA_3 and TIBA on growth, flowering and yield of different varieties of *Chrysanthemum. Ad. Plant Sci.* **16** (1): 143-146.

Mok, M.C. (1994). Cytokinins and Plant development: An overview. In cytokinins: Chemistry, Activity and Function, D.W.S Mok and M.C. Mok, eds (Boca Raton, FL: CRC Press), pp. 155-156.

Molina and Rowland, (1974). American society of plant biologists. Effects of ozone depletion on land plants. *Plant physiol.* **134:** 16-17.

Morie, E., T. Ozeki, K. Inove. M. Ishikawa and T. Tashiro, (1989). Rice flower glumes as an interceptor of UV-rays. *Jpn. J. Crop. Sci.* **58:** 540-548.

Munsi, S. K. and Kumari, A. (1994). Physiological characteristics of siliquae and lipid composition of seeds located at different positions in mature mustard inflorescence. *J. Sci. Food Agric.* **64:** 289-293.

Murali, N.S. and A.H. Teramura, (1985). Effects of UV-B irradiance on soybean VI. Influence of phosphorus nutrition on growth and flavonoid content. *Plant physiol.* **63:** 413-416.

Murphy A.S., Peer W.A. and Taiz L, (2000). Regulation of auxin transport by amino peptidases and endogenous flavonoids. *Planta* 211: 315-324.

Nagel, L., Brewster, R., Riedell, W.E. and Reese, R.N. (2001). Cytokinin regulation of flower and pod set in soybeans (*Glycine max* (L.) Merr.) Ann. Bot. **88:** 27-31.

Naidu, S.L., J.M. Sullivan, A.H. Teramura and E.H. Del Lucia, (1993). The effects of UV-B radiation on photosynthesis of different aged needles in field grown lobolly pine. *Tree Physiol.* **12:** 151-162.

Nanda, R., Bhargava, S. C. and Rawson, H. M. (1995). Effect of sowing dates on rate of leaf appearance, final leaf number and area in *Brassica compestris, B. juncea and B. napus. Field crop Res.* **42:** 125-134.

Nedunchezhian, N. K. Annamalinathanand and G. Kulandaivelu, (1995). UV-B (280-315 nm) enhanced radiation induced changes in chlorophyll, protein complex and polypeptide composition of chloroplasts in *Vigna unguiculata* seedlings grown at various temperatures. *Photosynthetica* **31:** 21-29.

Neeta Bhatt and Bhavtosh Bhatt, (2004). Effect of UV-B radiation on net primary productivity of *Lycopersicum esculentum*. Vol. **83:** 61-64.

Negash L. (1987). Wavelength-dependence of stomatal closure by UV-radiation in attached leaves of *Eragrostis tef:* action spectra under backgrounds of red and blue lights. *Plant physiologia and biochemistry* **25:** 753-760.

Noggle G.R. and G.T. Fritz. (1976). "Introductory plant physiology". Prentice Hall of India Priv. Ltd., New Delhi.

Normanly, J. (1997). Auxin metabolism, *Physiol. Plant.* **100:** 431-442.

Odum, E. P. (1960). Organic production and turn over in old-field succession. *Ecol.* **41:** 34-49.

Ohtani, Takeshi and Tadashi Kumagai, (1982). Phytochrome mediated effects of near UV radiation in the induction of flowering in etiolated *Lemna paucicostata* T–101, a shortday plant. *Planta* 153 (b): 543–546.

Pal, M. *et al.* (1999). *Plant physiol.* **4:** 79-84.

Pal. M. Jain V. and Sengupta, U. K. (1997). Influence of enhanced UV-B radiation on mustard cultivar response. *Indian J. Plant physiol.* **3:** 188-193.

Papadopoulos, Y.A., R. J. Gordon, K.B. McRae, R. S Bush, G. Belanger. E. A. Butlur, S.A.E. Fillmore and M. Morrison, (1999). Current and elevated leavels of UV-B radiation have few impacts on yields of perennial forage corps *Global change Biol.* **5:** 847-856.

Patil, A. A., Manipur, S. M. and Nalwadi, U. G.,(1987). Effects of GA and NAA on growth and yield of cabbage. *South Indian Hort.*, **35:** 393-394.

Peteropoulou, Y. A. Kyparissis, D. Nikolopoulus and Y. Manetas, (1995). Enhanced UV-B radiation alleviates the adverse effects of summer drought in two *Mediterranean pinus* under field conditions. *Physiologia Plantarum* **94:** 37-44.

Pharis, R.P. and King R.W. (1985). Gibberellins and reproductive development in seed plants. *Ann Rev Plant Physiology* **36:** 517- 568.

Phillips, I.D.J. (1972). Endogenous gibberellin transport and biosynthesis in relation to geotropic induction of excised sun-flower shoot-tip. *Planta* **105:** 234-224.

Prem Kumar, *A. et al.* (2001). *Plant Sci.* **161:**1-8.

Purkayastha, J., Sugla, T., Paul, A., Solleti, S., and Sahoo, L. (2008). Rapid in vitro multiplication and plant regeneration from nodal explants of *Andrographis Paniculata:* A variable medicinal plant *in vitro* cellular and development biology –Plant **44** (5): 442-447.

Purohit, S. S. (1988). Hormonal regulation of plant growth and development. Agro Botanical Publishers (India), Vol. IVth.

Pyle J. A. (1997). Global ozone depletion: observations and theory. In: Lumsden PJ, ed. *Plants and UV-B: responses to environmental change*. Cambridge University press, 3-11.

Raki Z. Johnson C.B. (2001). Influence of environmental factors (UV-B radiation) on the composition of essential oil of *Ocimum basilicum* (sweet basil). *Journal of herbs, spices and medicinal plants* **9** (in press).

Ram, K., Varma, A. N. and Sharma, P. K. (1973). Effect of NAA on growth, protein and ascorbic acid content of cabbage. *Plant science*. **5:** 150-153.

Rapoport, E. N., K. E. Heller, P. Dayanandan, F.V. Hebard and P.B. Kaufman, (1978). Role of indole-3 acetic acid and gibberellin in the control of internodes elongation in *Avena* stem sections. *Plant physiol*. **62:** 807-811.

Ravindran K.C. *et al.* (2001). *Biol. Plant.* **44:** 467-469.

Reis, J.M.R., Chalfum, N.N.J., Lima, Lc-De-O., L. C. (2000). Effect of etiolation and Indole Butyric Acid on the rooting of cuttings from the rootstock *Pyrus* (*Alleryana Dcne*). Ciencia-E-Agrotechnologia, **24:** 931-938.

Roa, H.K. and Kushwaha, H.S. (2005). Validation of CERES Rice model for prediction of upland rice yield. *J. of Agromet*. **7:** 101-106.

Ros, J. and M. Tevini. (1995). Interaction of UV-radiation and IAA during growth of seedlings and hypocotyl segments of sun-flower. *J. Plant physiol.* **146:** 245-302.

Rosa, T.M. De La. R.J. Tiitto, T. Lehto and P. J. (2001). Secondary metabolites and nutrients concentration in silver birch seedlings, under five levels of daily UV-B exposure and two relative nutrient addition rates. *New physiologist* **150:** 121-131.

Roychoudhary, R. and S. P. Sen. (1964). Studies on the mechanism of auxin action. *Physiol. Pl.* **17:** 352–362.

Russell, D.H., N. Bjorn and T. Elisabeth. (1998). Irradiance-induced alterations of growth and cytokinins in *Phaseolus vulgaris* seedlings. *Pl. Growth Regul.* **25:** 63-69.

Sadasivam, S. and M. Manickam. (1996). Biochemical methods (eds-2). New age international publication, New Delhi.

Sajjan, A.S., M. Shekhargouda and K.N. Pawar. (2003). Effect of regulatory chemicals on growth, yield attributes and seed yield of *Okra*. The *Orissa J. Hort. Sci.* **31** (2): 37-41.

Sandhu, M. S., R.G. and M. J. Kasperbauer. (1974). Localization of stem elongation control in *Cucumis sativas. Plant physiol.* **53:** 942-943.

Scannerini S. Maffei M. (2000). The effects of UV-B on photomorphogenesis and essential oil content in peppermint *(Mentha piperita). Journal of essential oil Research* **12:** 523-529.

Scantos *et al.* (2004). *Plant science.* **167:** 925-935.

Schumaker, M. M., J.H. Bassman, R. Robbereeht and G. K. Radamaker (1997). Growth, leaf anatomy and physiology of *Populus* clones in response to solar ultraviolet radiation. *Tree physiol.* **17:** 617- 626.

Sengupta *et al.* (1999). Changes in growth and photosynthesis of Mungbean induced by UV-B radiation. *Indian journal. Pla. Physiol.* Vol.4: 79-84. Division of plant physiology, (IARI) New Delhi.

Shah, S.H. and Samiullah. (2006). Effect of *Phytohormones* on growth and yield of Black cumin (*Nigella sativa* L.). Indian J. Plant *Physiol.* **11** (2: (N.S) 217-221.

Shamshery, A. P. and P. K. Gangwar. (1979). Sugar beet response to gibberellic acid (GA_3). *J. Indian bot. Soc.* **58**: Suppl. 79.

Shang, L.I, A.L., Dong, Y., Li, G.M., Han, J.M. and Wange, F. R. (2000). Effect of 6 BA on photosynthetic capacity in wheat flag leaf and 1000-grain weight under water stress. J. *Hebei Agri. Univ.* **23:** 20-24.

Sharma, S.K., Gosal, S.S. and Minocha, J.L. (1997). Effect of growth on siliqua and seed setting in interspecific crosses of *Brassica* Species. *Indian J. of Agricultural sci.* **67:** 166-167.

Sharma, M. M., R. Kumar, V. K. Jain and A. K., Goyal (1988). Some effects of UV-B irradiance on growth and development of pea seedlings. HCPB. **5:** 5-7.

Shein, T., and D.I. Jackson. (1971). Hormone interaction in apical dominance in *Phaseolus vulgaris Ann. Bot.* **35:** 555-564.

Shindell D.T., Rind D, Lonergan P. (1998). Increased polar stratospheric ozone losses and delayed eventual recovery owing to increasing green-house gas concentrations. *Nature* **392:** 589-92.

Shirley, Winkel, (2002). Current opinion in *plant Bios.* **5:** 218-223.

Singh Alka, Kumar Jitendera, Kumar Pushpendra. (2008). Effects of Plant growth regulators and sucrose on post harvest physiology, membrane stability and vase life of cut spikes of gladiolus. *Plant Growth Regul 55*: 221-229.

Skoog, F., and Miller, C.O. (1959). Chemical regulation of growth and organ formation in Plant tissue cultures *in vitro. Symp. Soc. Exp. Biol.* **11** :118-131.

Smith, J. L., D. J. Burritt and P. Bannister, (2000). Shoot dry weight, chlorophyll and UV-B absorbing compounds as indicators of plant sensitivity to UV-B radiation. *Annals of Botany* **86:** 1057-1063.

Sorg, (1996). *Stratospheric ozone* 1996, United Kingdom Stratospheric Ozone Review Group. Fifth report. London: HIMSO.

Srivastava, A.C. (1999). Changes in growth and photosynthesis of mungbean induced by UV-B radiation. *Indian journal. Plant physiol.* Vol. **4pp** 79-84.

Strid, A., W.S, Chow and J.M. Anderson. (1990). Effect of supplementary UV-B radiation on photosynthesis in *Pisum sativum. Biochemica et Biophysica Acta* **1020:** 260-268.

Sullivan, J.H. and A.H. Teramura. (1992). The effects of UV-B radiation on loblolly pine. 2. Growth of field grown seedlings. *Tree.* **6:** 115-120.

Sullivan, J.H. and A.H. Teramura, (1989). The effects of UV-B radiation on loblolly pine. 1. Growth, photosynthesis and pigments production in green-house grown seedlings. *Physiol. Plant.* **77:** 202-207.

Sullivan, J.H., B.W. Howells, C.T. Ruhland and T.A. Day, (1996). Changes in leaf expansion and epidermal screening effectiveness in *Pinus taeda* in response to UV-B radiation. *Plant physiol.* **98:** 349-357.

Sullivan; J.H., D.C. Gitz, M.S. Peak and A. J. McElrone. (2003). Response of three eastern tree species to supplemental UV-B radiation: Leaf chemistry and gas exchange. *Agri. For. Meterol.* **120:** 219-228.

Swaminathan, V. (1987). Response of Okra (*Abelmoschus Esculantus*). Moench to Growth Regulator treatments and foliar fertilization of nitrogen during Summer Season. M.Sc. Thesis, G.B. Pant Univ. Of Agri. And technology, Pantnagar (UP), Indian.

T.K. Van, L.A, Garrand and S.H. West. (1976). Affects UV-B radiation on net photosynthesis of some crop plants **16:** 715-718.

Taylor, R.M., O. Nikaido, B.R. Jordan, J. Rosamond, C.H. Bray and A. Tobin, (1996). UV-B induced DNA lesions and their removal in wheat *(Triticum aestivum L.* leaves. *Plant cell Envirunment*. **19:** 171-181.

Teramura A.H, Ziska L.H. (1996). UV-B radiation and photosynthesis. In: Baker NR, ed. Photosynthesis and environment. Kluwer Academic publishers, 435-450

Teramura, A.H. (1980). Effects of UV-B irradiances on soybean. Importance of photosynthetically active radiation in evaluating UV-B irradiances effects on soybean and wheat growth. *Physiologia plantarum* **48:** 333-339.

Teramura, A.H., M. Tevini and W. Iwanzik. (1983). Effects of UV-B radiation on plats during mild water stress. Effect on diurnal stomatal resistance *plant physiol.* **57:** 175-180.

Teramura, A.H., Sullivan, J.H. (1994). Effect of UV-B radiation on photosynthesis and growth of terrestrial plants. *Photosynth. Res.* **39:** 463-473.

Tevini M. and A.H. Teramura, (1989). UV-B effects on terrestrial plants. *Photochem. Photobiology.* **50:** 479-487.

Tevini, M. (2000). UV-B effects on plant. PP 83-97, In: S.B Agarwal and M. Agarwal. Environmental pollution and plant Responses, Lewis Publishers, Boca Raton. USA.

Tevini, M., W. Iwanzik and U. Thomas, (1981). Some effects of enhanced UV–B irradiation on the growth and composition of plants. *Planta* **153:** 388–394.

Tevini, M., J. Braun, G. Fieser, U. Mark, J. Rao and M. Saile, (1990). Effects of enhanced solar and artificial UV-B radiation on growth, function and composition of crop plant seedlings. BPT–Final Report GSF Munchen.

Thimann, K.V. (1980). The senescence of leaves. In: senescence in plants. K.V. thimann (ed.) CRC press Inc., Florida. 85-115.

Thinann, K.V. (1997). Hormone action in the whole life of plants. University of Massachusetts Prens, Amherst. 448pp.

Thomas A. Day and Patrick J. Beale. (2002). Effects of UV-B radiation on terrestrial and aquatic primary producers vol. **33:** 371-396.

Vaithialingam, R. and J.S. Rao. (1973). Pre-soaking effects of maleic hydrazide (MH-30) on Groundnut. Madras Agric. J. **60**: 404-405.

Vu. C.V., Allen. L.H. and Garrard, L.A. (1981). Effect of supplementary UV-B radiation on growth and photosynthesis of soybean. *Physiol. Plant.* 52: 353-362.

Vu, C. V., L. H. Allen and L. A. Garrard, (1982). Effects of UV–B radiation (280–320 nm) on photosynthetic constituents and processes in expanding leaves of soybean [*Glycine max* (L) Merr.] *Environ. Exp. Bot.* **22** : 465 – 473.

Waithaka K. Dodge L.L. and Reid S.S. (2001). Carbohydrate traffic during opening of gladiolus florets. *J. Hortic. Sci. Biotechnol* **76:** 120-125.

Webb, A.R. (1997). Monitoring changes in UV-B radiation. In Lumsden PJ, ed. *Plants and UV-B: responses to environmental change.* Cambridge University Press, 13-30.

Wellmann, E. (1982). Phenylpropanoid pigment synthesis and growth reduction as adaptive reactions to increased radiation. **5/82:**145-149.

White J. C.G.C. Medlow, J.R. Hillman and M.B. Wilkins, (1975). Correlative inhibition of lateral bud-growth in *Phaseolus-vulgaris.* Isolation of IAA from the inhibitory region. *J. Exp. Bot.* **26:** 419-424.

Wilson, M.I., Greenberg, B.M. (1993). Specificity and photomorphogenic nature of UV-B induced cotyledon curling in *Brassica napus* L. *Plant physicl.* **102:** 671-677.

WMO, (1995). *Scientific assessment of ozone depletion;* (1994). World Metrological Organization, Global Ozone Research and Monitoring Project, Report No. 37, Geneva.

Wolbang Carla M., Peter M, Chandler, Jennifer J. Smith and John J. Ross. (2004). Auxin from the development inflorescence is required for the biosynthesis of active Gibberellins in Barley stems. *Plant Physiol.* **134:** 769-776.

Wort, D. J. (1964). Effects of herbicides on plant composition and metabolism. In: "The physiology and biochemistry of herbicides". Ed. L. J. Audus Academic Press, New York. pp. 291–334.

X. Hao, B. A. Hale, D.P. Ormrod, A.P. Papadopoulos. (2000). Effects of pre-exposure to UV-B radiation on responses of tomato (*Lycopersicon esculentum*) to ozone in ambient and elevated CO_2. *Environ. Pollu.* **110:** 217-224.

Yadav, P.I. Poornima and P. Babu Mathew. (2005). Plant growth promoters for sustainable rice (*Oryza sativa* L.) Production. Indian J. Bot. Res. 1 (1): 73-78.

Yang, Y., Y. Yao, X. Gang and C. Li. (2005). Growth and physiological response to drought and elevated UV-B in two contrasting populations of *Hippophae rhamnoides. Plant physiol.* **124:** 431-440.

Yan Shengrong, Huang X. and Zhou Qing. (2007). Effect of Lanthsnum on the oxygen metabolism of soybean seedlings, under supplemental UV-B irradiation.

Yatsuhashi, H. and T. Hashimoto. (1985). Multiplicative action of a UV–B photoreceptor and phytochrome in anthocyanin synthesis. *Photochem. Photobiol.* **41**: 613 – 680.

Yim, K.O, Known, Y.W., Bayer, D.E. (1997). Growth responses and allocation of assimilates of rice seedlings by paclobutrazol and gibberellin treatment. *J. Plant Growth Regul.* **16:** 35-41.

Yu, J. (1999). Parthenocarphy Induced by N- (2-Chloro-4-Pyridyl) – N'- Phenylurea (CPPU) prevents flower abortion in Chinese White-flowered gourd (*Lagenaria leucantha*) *Environ. Exp. Bot.* **42:**121-128.

Yuan, L. Z. Yangum, C. Aryan and H. Zhide, (2000). Intraspecific responses in crop growth and yield of 20 wheat cultivars to enhanced UV-B radiation under field conditions. *Field Crops Res.* **67:** 25-33.

Yuda E, Matsui H, Nakagawa S, Yukimoto M. and Wada K. (1984). Effect of 15-β-OH gibberellins on the fruit set and the development of three pear species. *J Jpn Soc. Hort. Sci.* **53:** 235-241.

Yue, M., Y. Li and X. Wang, (1998). Effects of enhanced UV-B radiation on plant nutrients and decomposition of spring wheat under field conditions. *Environmental and Experimental Botany* **40:** 187-196.

Zavala J. Scopel A.L, Ballare, C.L. (2001). Effects of solar UV-B radiation on soybean crops.

Zhang Z, Zhou W, L. I. H. (2005). The role of GA, IAA and BAP in the regulation of in vitro shoot growth and micro tuberization in potato. *Acta Physiol Plant* **27:**317-323.

Zhang C., Tanabe K., Tamura F., Itai A. and Yoshida M. (2007). Role of gibberellins in increasing sink demand in Japanese pear fruit during rapid fruit growth. *Plant Growth Regul.* **52:** 161-172.

Index